Containing Nature

John Coupland

Containing Nature

The Ongoing Invention of Canned Food

John Coupland
Department of Food Science
The Pennsylvania State University
University Park, PA, USA

ISBN 978-3-032-11881-3 ISBN 978-3-032-11882-0 (eBook)
https://doi.org/10.1007/978-3-032-11882-0

This Springer imprint is published by the registered company Springer Nature Switzerland AG
The registered company address is: Gewerbestrasse 11, 6330 Cham, Switzerland

To my parents: Neil and Lorna

It is very easy to talk learnedly and at length about science in the abstract, but unless one can show just how this science has a practical bearing on the problems that are daily encountered in the industrial world, such talk is vain and may be wrongly construed.

Samuel Prescott in a talk to the Atlantic States Packers' Association and the Western Packers' Canned Goods Association, Detroit, Michigan, 1899

We strongly advise all parties going into the canning business to secure at the start a man of practical experience to superintend the processing of the goods

Sprague Canning Machinery Co. Catalog, 1903

Preface

I arrived at my interview for a position in the Department of Food Science at Penn State feeling reasonably prepared. I had a bachelor's degree and a PhD in food science. I'd written a handful of papers that I was keen to talk about. At first, all seemed to go well. I was ushered through different labs and offices where I shared my ideas with students and faculty. Their questions were thoughtful, their smiles encouraging. Then, that afternoon, they took me to a vegetable canning factory.

The day of my interview coincided with a meeting of the local branch of our professional society, the Institute of Food Technologists (IFT). It involved the tour of a local family-owned business followed by dinner and a talk from the patriarch, who emphasized the importance of Christian faith, family, and sheer persistence in achieving success. At the time, most members of the department would attend these meetings, so they thought it might be nice to invite the candidate along too.

This all makes sense now, but at the time I was trapped in the low-grade paranoia that often goes with job interviews. *This is a test*, I thought. *Of course it's a test. But of what? What could I possibly say that would matter to the people who run this factory?* Not much, as it turned out.

There I was, in a real food factory, and despite my doctoral degree, interview suit, and academic papers, I had little to contribute. Sure, I had vague memories of how canning worked in theory from my undergraduate courses, but the people here had real expertise. They received truckloads of vegetables, metal cans, and labels at one end of the plant and shipped out truckloads of packaged food at the other. They had huge pieces of complex equipment—some new and computerized, others obviously in use for decades—clunking and giving off belches of steam. They hired skilled and unskilled workers. They worked with regulatory authorities, farmers, accountants, and supermarket buyers. They made food that people wanted to eat, they made a profit, and so paid the wages that supported lives.

Despite my efforts to be "applied," there was a huge gap between the things I knew and the things they did. A gap my education had not prepared me for. Thankfully the visit wasn't a test, and I was lucky enough to get the job and embark on a career of teaching and academic research.

Many years later, and perhaps slightly less naive, I returned to the Institute of Food Technologists and served a term as its president. In that role, I had the opportunity to meet people with various types of technical expertise in foods. I met people who make bourbon and people who make cola, people who blended flavors and people who cultured bacteria, people who grew strawberries and people who fed astronauts. I encountered chemists, engineers, microbiologists, statisticians, nutritionists, physicists, and teachers—although most of them identified themselves as food scientists. I met interns, entrepreneurs, executives, and retirees. I met the man who invented Tang™. They worked in offices and labs, farms and factories. They were employed by companies, governments, universities, and associations. They weren't connected by any single set of skills, job functions, or disciplines, but they were all valued for their expertise in some aspect of making food, at scale, for profit.

There are a lot of popular books about food science and the food industry, but most of it has a strong moral aspect. Authors are keen to explain either how modern industrial food is poisoning the world and making us all sick, or how innovation is the only way to feed the growing billions. However, there is much less written about the work done to make everyday foods possible. Certainly, the food industry itself rarely seeks to publicize it. They recognize that the words "science" or "technology" or "factory" do not stimulate the appetite, so they instead often focus on romantic images of farms, chefs, and kitchens.

But there are scholars interested engaged with this sort of thing: historians, sociologists, and philosophers interested in science and technology both in general terms and, often through the multidisciplinary field of food studies, in food in particular. These perspectives could help food scientists understand their fields, but they are often quite specialized and not widely read outside their disciplines.

My approach has been to use history to provide examples of people working with the same basic technology in different ways at different times. I have drawn on the work of scholars in the humanities and the social sciences to find theories which provide context. I recognize there is risk in stepping outside my own discipline, but I have tried to confine myself to "translating" the work I cite into a form that is more relevant to working scientists and technologists. What I have tried to add is from my own experience doing science, and from the interviews I have been able to conduct with people working in the industry.

University Park, PA, USA John Coupland

Acknowledgments

At the risk of a “spoiler,” one of my conclusions is that progress is a social process; nothing useful is created in isolation. This has been my experience in writing this book. To explore the questions I am interested in, I have had to travel far from my academic roots, and the process of meeting new people—either through their writing or in conversation—has made the journey rewarding. Without their generosity, this would not have been possible.

I grateful to all the friends; colleagues and strangers who offered general guidance and conversation; specific information on historic details or modern practice; or read and critiqued drafts. In particular: Nathan Anderson; Claire Berton-Carabin; Lorna, Jennifer, Michael and Anna Coupland; Colin Dennis; Chris Findlay; Frank Furman; Tim Haley; John Hayes; Scott Hearn; Keith Ito; Josh Lambert; Rachel Laudan; John Litchfield; Tony Pavel; Nathan Pearlman; Gregg Pearson; Phil Perkins; Jonathan Rees; Ken Reid; Philip Richardson; Bob Roberts; Laura Rolon; Charles Santerre; Julia Schmidt; John Spevacek; Rick Stier; Nikolas Stoforos; Herbert Stone; Malcolm Summers; Christie Tarantino-Dean; Arthur Teixeira; Don Thompson; Stiphany Tieu; and Anna Zielde.

I am also grateful to my colleagues at Penn State for their support. This book began as a COVID project and evolved into a sabbatical project and dragged on from there—as these things often do. It took me away from the types of scholarship typically expected in my role as a professor of food science. Still, the privilege of a tenured position demands you take some risks along the way. I am very lucky that my interview all those years ago worked out the way it did and gave me the opportunity to try.

Most of all, I offer my deepest thanks to my family for their constant love and support through all the ups and downs of this long effort. When my dinner conversations turned, once again, to microbiological methods in the 1920s or Alaskan canneries in the 1900s, or microscopy in the seventeenth century, you made a splendid effort to remain interested and helped me believe other people might be interested too.

Acknowledgements

Contents

Chapter 1
All the Science You Don't See

Abstract The way we use technology reflects the way we live our lives. Much of the technology we depend on is "invisible" to us and this has costs for society in general and for technology workers in particular. This book will tell stories from the history of canned food to illustrate the nature of technology work and how it forms our lives.

1.1 Canning as an Invisible Technology

In 1949, Andrew Warhola left Pittsburgh for New York and became Andy Warhol. He quickly found success drawing consumer products for magazine ads but longed to establish himself as an artist. His breakthrough came in 1962 with a striking exhibition of paintings—Warhol called them "portraits"—depicting all the current varieties of Campbell's canned soup. The work was controversial, and the controversy made him famous.

There are many explanations for why Warhol picked canned soup as his subject. One friend had suggested he should paint "something people see every day" because the impact of pop art depended on the viewer having an existing experience of the subject [1]. Warhol himself said simply "Because I used to drink it. I used to have the same lunch every day, for twenty years, I guess, the same thing over and over again" [2]. Campbell's was the biggest selling brand of soup in America, but, although their cans were seen by millions of people every day, they were never really *seen* until Warhol put them on the wall and called them art. The cans were the ideal subjects because, by 1960 in America, they were already mundane.

If the cans themselves were mundane, the processes that brought them to the supermarket shelves were completely invisible. There were systems of contracts for farmers to deliver specified grades of vegetables to the factory at specified times, there were automated machines to first make and then fill the cans, there were scientifically based rules for how long the cans had to be cooked and there were training and validation systems to ensure these rules were applied, there were storage and distribution networks, and there was advertising. Very little was secret, but very little was seen.

J. Coupland, *Containing Nature*, https://doi.org/10.1007/978-3-032-11882-0_1

According to Arthur C. Clark, "any sufficiently advanced technology is indistinguishable from sorcery" [3]. However, perhaps the most magical thing a truly advanced technology can do is to simply disappear. As the historian of science Steven Shapin has noted, we are surrounded by all kinds of "embedded science" which we depend upon yet remain completely unconscious of [4]. Much of the disappearing trick is simply human attention bias. We notice new things as "technology," but, with sufficient familiarity, things become part of the background.

You can start to bring this invisible world of technology into focus with a simple experiment. If I asked you to list the most important technologies of today, you might include generative AI, gene editing, wireless communications, robots, rockets, or even a popular website. These are not necessarily the newest discoveries, as those are likely only of interest to the most specialized research communities. What we identify as "important" are things which have developed to a point that they are doing something new in our lives or, even more likely, we believe that they might do something soon. They capture our interest because they represent change or the potential for change.

Now if I asked you instead to list the technologies you are depending on right now, your list would be different. You might start slowly, as you begin to realize what would immediately threaten life or livelihood if they disappeared—clean water and electricity supplied to your home, a cheap generic medication, convenient transportation—but the list never stops. Think about the number of mundane objects you have touched today: the pencil you write with, the buttons that fasten your clothes, the building materials that make up your home, breakfast cereal, toothpaste. These might not be matters of life and death, but your day would be noticeably worse without them. Now think more deeply and consider the objects that made the objects you touched today: the machines that made the plastic into water pipes and buttons and toothbrushes. Think about the machines that formed that plastic out of oil. Think about all the manufactured parts of all those machines. Slowly, an enormous but previously unseen world of human effort comes into focus.

Our complete dependence on the things we make is one of the defining features that sets us apart from other species. An animal depends on its body to find the food and shelter it needs to live. An eagle's talons, a bear's thick fur, and the shape of a finch's beak all allow their owners to succeed in doing specific things in their specific environments. Humans are different. We use our bodies to make things and then use these things to create what Ursula Le Guin called an "active human interface with the material world" [5]. Adapting our tools, our interface, allows us to do many different things in different environments. The things we make and the ways we use them shape our lives.

Along with our dependence on made things comes another distinctively human characteristic—our dependence on one another. From Robinson Crusoe to Bear Grylls, we have loved stories of people surviving alone in a harsh wilderness, but these stories are merely fantasies. In reality, these brave individuals only survive because of the things they bring with them—things that other people made. People thrive in groups. It is the set of technologies the group can together bring to bear that provides each individual with their interface with the natural world.

The story of humanity is therefore about our use of material objects, our technology, but also our relationships with one another in societies. It is not surprising that, throughout human history, changes in technology use and changes in social organization are closely related. This is most immediately evident in the ways we feed ourselves.

1.2 How Technology Shaped Humanity

Humans have depended on their technology long before *Homo sapiens* first appeared on the planet. The anthropologist Richard Wrangham argued that humans were able to survive with smaller jaws, shorter intestines, and bigger brains than other primates because we used the technology of fire to cook our food and make it both softer and more digestible [6]. Cooked food was a valuable resource accumulated in one place, so larger social groupings became valuable to either protect or to steal it [7]. Our ancestors used fire and other technologies as "interfaces" to make the material world more to their liking, and as a result their bodies and their societies changed to become more like ours, more characteristically human.

Beginning around 12,000 years ago, humans began a slow but profound transition in the ways they fed themselves—shifting from hunting and gathering to farming and raising livestock. This transition, known as the Neolithic Revolution, introduced new ways of interacting with the natural world and a new suite of technologies: digging sticks and plows for cultivation, granaries for food storage, millstones for grinding grains, and innovations in animal husbandry and irrigation.

People used the new tools to modify their environment and made certain places—such as a river valley where grains grew well—more valuable. They concentrated around these places, where they developed larger and more hierarchical societies [8]. These societies required their own set of technologies: writing and numeracy to track the ownership of grain, new weapons to protect or raid food stores as well as to enforce the new social order, and decorative arts and monumental architecture as status symbols and to support the religious practices that bound societies together. The people born after the Neolithic Revolution were no more intelligent than their ancestors, but they lived radically different lives. Their tools and technologies became a new kind of "active human interface with the material world," reshaping not only how they survived, but how they understood and organized their world.

The second decisive change in humanity's relationship with technology began in northern England in the second half of the eighteenth century—the Industrial Revolution [9]. Starting with textile manufacturing, people learned to make things using powered machines to replace human labor. Production shifted to large factories centered around a power source—water or later steam—and people followed the work into growing industrial cities. The pattern spread to other industries and other places, and, by the middle of the nineteenth century, much of Europe and North America was industrialized. A second wave of industrialization based around interchangeable parts made by machine tools, electrical power, the chemical industry, and the internal combustion engine developed in Germany and America in the second half of the nineteenth century and again spread around the world [10]. Some scholars have argued the increased use of digital computers in the second half of the twentieth century constitutes a distinct third Industrial Revolution and claims of a "fourth Industrial Revolution" are frequently attached to the latest business fad.

The adoption of industrial technologies accelerated economic growth, as the same amount of human effort produced more products. Material goods became cheaper and more widely available. People moved from rural villages, where they worked as farm and craft laborers, to larger, more diverse cities where they worked in factories and struggled for political power in new ways. Distances and time were compressed by advances in transportation and telecommunications. Natural resources were exploited more rapidly, and landscapes radically transformed. Pollution from factories and sewers sickened people, yet at the same time both lifespans and populations increased. The people born after the Industrial Revolution were no more intelligent than their ancestors, but they lived radically different lives. Their tools and technologies became a new kind of "active human interface with the material world," reshaping not only how they survived but how they understood and organized their world.

Of course, there had been technological advances before the Industrial Revolution, but these had frequently been met with caution because of the social disruption they brought with them. New technology that threatened people's livelihoods was often resisted by workers who organized politically and even wrecked machines. New technology that threatened the existing social order was also resisted by elites who feared their own power would be undermined. One of the reasons technological change took hold in eighteenth century England is that the ruling class had enough financial interests in the new factories to side with their owners and use the force of law to suppress worker resistance [11]. The industrialists and political elite grew rich and, even if the initial stages of the Industrial Revolution were devastating to the health and wealth of working people, the new technologies continued to be adopted and built upon one another.

Indeed, one of the features that distinguished the Industrial Revolution from earlier phases of technological development was an increasing pace of change. Successes in one area supported further technological innovation in another. For example, improvements in steam engines led to advances in shipping, mining, industrial manufacturing, and railroads, while the growth of these industries in turn increased the demand for steel production. Improvements in transportation, particularly railroads and steamships, meant it was worth investing in specialized factories that could be connected to natural resources and markets thousands of miles away. People learned to see their world changed by new technologies many times over the course of their lifetimes.

Before the Industrial Revolution, people believed religious, commercial, or political change were the only ways a society could be improved. If we could reform our church or get the rights to trade in that city or win this war, then our lives would be better. It was only after the Industrial Revolution that the adoption of new technologies regularly changed people's lives. If we can prevent management from replacing our jobs with robots or build a new water treatment facility for our town or install some new software to let distant family members join a group chat, then our lives will be better. This is, perhaps, the main reason for our modern attention bias toward new technologies. People have always used technology, but it is only after the Industrial Revolution that have we come to expect our lives to be transformed by technological advances. Anything that offers such a promise of risk and reward is worth paying close attention to.

1.3 Technologists and Scientists

We can get another perspective of the ways technology shapes humanity by looking at the changing ways people have worked with it. Imagine three situations—an individual isolated from the world, a remote rural village, and a growing industrial city—and how people in those places might work with technology.

First, consider people separated from the outside world and depending only on the things they can make and use to provide the necessities of life. Examples include the Jack London's foolhardy traveler in the Yukon in *To Build a Fire* or the contestants on the *Naked and Afraid* TV show. They must maintain a fire, find shelter, and gather food to survive. They fail because they lack both the technology skills and the connection to a wider society that could support one another. They are cautionary tales.

Next, imagine a small, remote peasant village. Fictional examples include *Things Fall Apart* by Chinua Achebe, *To Live* by Yu Hua, and several novels by Wendell Berry. Most people are farmers, with deep expertise in growing crops and raising animals which they learned within their families. Depending on their environment and history, they might know how to store and plant seed, how to treat animal diseases and breed better livestock, or how to process their harvest into edible foods. In a functioning society, people can always depend on relationships to share resources, so not everyone needs to perform every task. Some household members might work in the fields, while others grind grains or make bread. Still, most adults would either be directly responsible for the technologies they depended on or closely connected to those who were. The few tools or goods too complex to produce at home would be acquired through trade, using surplus farm products or labor as currency.

Exchange relationships like these allowed some people who were particularly good with specialized technologies—artisans such as millers, potters, and blacksmiths—to focus on making objects and trading them for their food and shelter. In smaller villages, the opportunities for specialized technical work were limited; a farmer might have a side business making pots or repairing tools. But in larger cities, specialists could refine their skills and focus more narrowly on their craft. Ancient Rome, for example, had engineers specialized on different types of building project. The artisans would learn their trades through apprenticeships, or within families, and in some cases tried to protect their valuable knowledge by forming trade guilds.

For our third example of people working with technology, imagine a growing city in northern England in the early of the nineteenth century. Fictional examples include *North and South* or *Mary Barton* by Elizabeth Gaskell or *Hard Times* by Charles Dickens—all describing a fictionalized version of Manchester. The Industrial Revolution had turned many poor, rural farm workers into poor, urban factory workers. The old way of manufacturing things required skilled labor, and the artisans could sometimes demand good wages. In contrast, industrial factories relied on machines to mass-produce goods, and most jobs involved repetitive tasks to keep those machines running. Very few adults directly used the technology they needed for their food and shelter, nor did they know the people who did. Instead, workers fed themselves by spending their wages on foods grown far away and often processed in other factories by strangers.

A second, smaller group of technology jobs developed following the Industrial Revolution: those responsible for building and maintaining the new machines. At first, the people taking these roles had learned an artisanal skill and had the ingenuity, persistence, and capital to apply it to the growing industrial opportunities. For example, the key inventions that led to the first industrial textile factories in eighteenth century England (James Hargraves' spinning jenny in 1764, Richard Arkwright's water frame in 1769, and Samuel Compton's spinning mule in 1779) were all first built by men from backgrounds in clothmaking and related trades but with almost no formal education. Later though, more specialized skills were needed to work with the increasingly precise machines, and the types of people who succeeded were more educated. The training of these engineers became more formalized as the nineteenth century progressed, with specialized schools teaching mathematics and drafting alongside the craft skills [12].

Later still, a third type of technology worker, perhaps the smallest group of all, also found its place in the new industrial economy. European science had begun to emerge from natural philosophy in the "Age of Enlightenment" of the seventeenth and eighteenth centuries. The new scientific discoveries were interesting, but not immediately relevant to the first stages of the Industrial Revolution. By the second Industrial Revolution, however, advances in chemistry and electricity were economically essential. Universities, especially those in Germany, responded by training students in the applications of their subjects to industrial problems and their graduates found work in factories.

There was usually some level of class distinction between these groups of industrial workers, with the "unskilled" machine operators at the bottom, the educated scientists in their white coats at the top, and the engineers somewhere in between.[1] The distinctions were blurred: was the person you could rely on to get an old machine running or make a replacement part really an engineer? Was someone who tested different formulations for the filament in an incandescent lamp or new recipes

[1] A widely repeated anecdote is that when James Alfred Ewing was appointed as the Professor of Mechanism and Applied Mechanics at Cambridge University in 1890—their first engineering position—his colleagues sarcastically suggested that the university might as well hire a "Professor of Jam-Making."

for canned soup really a scientist? Sometimes new titles were created to cover this space—a mechanic, a technician, a technologist—but using these carried class implications too. For someone who considered themselves a scientist to be called a technologist diminished their status, while for someone who considered themselves a factory worker, it was a promotion.

It is primarily scientists and engineers—those in the last two specialist groups—who have driven the technological changes that have transformed life since the Industrial Revolution. These groups are both the subject of this book and its intended audience. The narratives we construct about their work are often shaped by two biases: a cultural preference for new technologies over established ones and a social preference for science over engineering. These biases, I believe, most significantly disadvantage the technology workers themselves. We can see these dynamics at play in the history of canning and in how modern food technologists have interpreted that history.

1.4 Who Remembers Nicolas Appert?

Canning was invented in early nineteenth-century Paris by Nicolas Appert [13]. Appert is remembered in France with plaques and postage stamps, but he is less well known in the English-speaking world than the other famous inventors of the period—Eli Whitney and his cotton gin or George Stevenson and his steam locomotive. The one group outside France that continues to celebrate Appert today, perhaps almost as a secular "patron saint," are the scientists and engineers who work on food. Among the first members of the cult of Appert was Katherine Golden Bitting.

Katherine Bitting trained as a botanist at Purdue University before working on food preservation at the USDA and the National Canners Association alongside her husband, Arvill. She lived through a time when the Industrial Revolution swept through American food production. Over the course of her career, she saw science and technology used to transform the quality, safety, and variety of the foods that people could afford, and she thought that was wonderful. Bitting became fascinated by the story of this progress and with its foundation in canning, the first modern processed food.

In 1920, Katherine Bitting published her English translation of Nicolas Appert's 1810 book on canning. Arvill Bitting republished this in 1937, along with other materials relevant to the early history of canning, in a book *Appertizing*. The title was a pun, but also a term they felt should be used in place of canning to better recognize Appert's pioneering role. The opening chapter was an essay by Katherine entitled *A Benefactor of Humanity* describing Appert's life and achievements in glowing terms.[2] She had no doubt that Appert, and by extension food science and

[2]The title "Benefactor of Humanity" was taken from an award given to Appert late in his life by the French government. Bitting's essay was reprinted and developed from a collection of materials she had used in canning courses offered by a canning trade association since the early years of the

technology generally, had benefited humanity enormously by improving nutrition and making delicious foods available out of season. This was the message that continues to appeal to food scientists and technologists as a group and why they have maintained a reverence for Appert and his work. In 1942, after Katherine's death, Arvill Bitting was instrumental in naming the most prestigious award available to a food scientist for Nicolas Appert [14]. The list of annual awardees is a catalog of the triumphs of the profession [15].

Ever since then, working food scientists have returned to the history of canning and particularly to the work of Appert, sometimes adding additional scraps of research but often just retelling the same essential narrative. In 1957, *Food Technology* magazine reported some research clarifying Appert's first name and described him as "creator of the food preservation industry" [14]. In 1975, another paper in the same journal revealed details of his baptism and said he was "responsible for the creation of the canning industry" [16]. Food scientists Graham [17] (in 1981), Farrer [18, 19] (in 2004 and 2008), Garcia and Adrian [20] (in 2009), and Featherstone [21] (in 2018) wrote historic accounts of Appert's life in science journals. Stuart Thorne studied the science of canning for his doctorate, and, while teaching the subject at the University of London, wrote *The History of Food Preservation* with a strong focus on the history of canning [22]. Norman Cowell was a physicist who did research work on heat transfer in foods and who, late in his career, earned a doctorate in food science with a study on the early history of canning [23]. Cowell published his work, not in history journals, but once again in *Food Technology* magazine for an audience of food technologists [24, 25].

Clearly the story of Appert remains interesting to people doing technology work in foods. However, by focusing their interest on Appert the inventor, food technologists are submitting to the widespread misconception that technology is only important when it is new. In doing so, they were missing the value their own work.

Appert's first canned foods were hand-packed into glass bottles and sealed with corks. The bottles were boiled for however long the canner thought prudent, who then watched carefully to see if they exploded on storage or stank after they were opened. They were only available in select shops, and they were expensive. The few people that tasted them at the time were impressed, but they were very few. A hundred and 50 years later, Warhol's mundane cans came streaming off the Campbell's production lines like endless gleaming robotic snakes. The processes would have been beyond Appert's imagination. Between the invention of canning and Warhol, there had been countless small changes which transformed the crude, hand-filled bottles to fully industrialized products that were cheaper, safer, and a common part of people's diets.

Nicolas Appert certainly invented canning, but who invented Andy Warhol's mundane cans? Pretty much everyone who had done technical work on canning over the intervening years to make the process better. When we choose to talk about

century. Much of the historic material from *Appertizing* was republished as late as 1961 in canning textbooks showing a long-standing interest in the story [26].

technology exclusively in terms of the inventors, we are neglecting all their achievements. Ironically, the people most prone to this error are the group most interested in Appert today, the food scientists and technologists who do this work.

We can get a richer understanding of technology and technology work by looking not only at the moment of invention but also as it matures. In doing so we can better appreciate the contributions of the nameless workers who turned a novelty into something that is such a part of our lives that we are barely aware of it as technology. This is a more egalitarian perspective, and one I think is particularly important to students and people working in technology today as it gives value to their contributions.

1.5 Retelling the Story of Canned Food

I chose canned food because it has a relatively well-described history with a clear invention date near the start of the Industrial Revolution and whose evolution spans many of the changes associated with industrialization. Canning is also an important technology in the foundation of my own field of expertise, food science. Not only because a lot of canned foods are eaten but also because many of the institutions in the field—academic programs, technical journals, professional societies—were first organized by the canners. As we have seen, food scientists have frequently turned to the example of canning to tell the story of their profession.

However, what we learn from canning may serve as an example for all the other established and invisible technologies we rely on every day, and so this book may be interesting to technologists in general and not just food technologists.

I have divided the development of canned foods into two main parts. In the first part, we follow selected developments in the technology of canned food: the machines and the materials and the processes. The characters in this part of the story are all engaged in the business of food preservation. Their work was rarely centrally planned but represents an evolutionary development based on many individuals looking at their work and thinking "that could be better."

We start in Nicolas Appert's Paris workshop just after the French Revolution (Chap. 2). Here we will see the invention of canning had no "eureka moment" but was a struggle of decades to assemble existing things in new ways to serve a new purpose. Despite this, we will end up defending Appert's reputation on the grounds that he was copied. We next follow cans on their murky path across the English Channel to a factory in South London where entrepreneurs used their superior technical backgrounds and a better commercial environment to make money on Appert's idea (Chap. 3). These will be our first examples of what modern technology work looks like, and I will argue adaptation and improvement are more important and even more rational than invention. Finally, we see that process at work not just in one place but in many small workshops across America (Chap. 4). In the century before the First World War, the combined efforts of many innovators transformed canned food from something that was expensive, handcrafted, and novel to something that was cheap, machine made, and mundane.

Next, we will use two examples of specific foods being canned in specific places—salmon in the Pacific Northwest in the late nineteenth century (Chap. 5) and Campbell's soup in Camden, New Jersey, in the early twentieth century (Chap. 6)—to complicate the simple "bigger, faster, more efficient" narrative of industrialization. Technology might reduce the need for labor but never completely replaces it. The workers are not robots, but people who have preferences about how the work should be done. The foods being canned are not some standardized materials fed into the machine but instead are variable, seasonal, and prone to spoilage. Successful factories depend on human judgments about how technology (machines), labor (people), and nature (the food) can work best together in a particular place and time.

In the second part of the book, we follow the parallel development of the science of canning. This is work done by people with significant training in science and with some social and economic distance from the practical business of canning. In this story, the changing nature of the relationships between scientists and canners is as important as the advances in the science itself.

At the start of our story, science and technology are not at all connected. The scientific ideas to explain how canning worked were established for at least 150 years before Appert's invention, but Appert was unaware of them. Indeed, when scientists were called on by the French government to explain Appert's achievement, their explanation was wrong, yet their error made no real impact on the development of the industry (Chap. 7). It was only at the end of the nineteenth century when developments in microbiology and changes in the ways American universities were organized led academic researchers to work closely with canners to come up with results that were practically useful (Chap. 8).

The next advance in canning science occurred soon afterward but first required another change in social organization. Canners came to realize that they could cooperate on the issues that affected them as an industry, while still competing against one another as businesses. They formed trade associations able to address the bad publicity from deadly botulism outbreaks which was damaging the reputation of all canned foods. The National Canners Association supported collaborative research and established scientific standards for canning which formed a basis for regulations that ensured canned products would be safe—whether or not they were made by Association members (Chap. 9).

The final story in the science of canning describes a situation when communities of scientists disagreed with one another. Environmental scientists used methods from endocrinology to argue that a component in can linings, BPA, is harmful. Traditional toxicologists, notably those supported by the industries that make and use BPA, disagreed. Both sides tried to use the media to influence public opinion, while the canning industry tried quietly to back away from controversy by using other materials to line cans (Chap. 10).

While the chapters in each section—technology, examples, science—are broadly in historic order, this is not intended to be a comprehensive history of canning. Instead, each chapter uses a detailed vignette to illustrate larger points about science and technology work. I close each chapter with a short reflection on how these might apply to people working in applied science and technology today. While the different chapters build upon one another to an extent, they can easily be read out of sequence according to interest.

References

1. Belz, C.: Pop art and the American experience. Chic. Rev. **17**, 104–115 (1964)
2. Swenson, G.: What is pop art? Answers from 8 painters, part 1. ArtNews. **62**, 25–27 (1963)
3. Clarke, A.C.: Clarke's third law on UFO's. Science (80-). **159**, 255 (1968)
4. Shapin, S.: Invisible science. Hedgehog Rev. **18**, 1–9 (2016)
5. Le Guin, U. K.: A rant about 'technology'. Available at: https://web.archive.org/web/20200523131531/; https://www.ursulakleguinarchive.com/Note-Technology.html (2004)
6. Wrangham, R.W.: Catching Fire: How Cooking Made Us Human. Basic Books (2010)
7. Wrangham, R.W., Jones, J.H., Laden, G., Pilbeam, D., Conklin-Brittain, N.: The raw and the stolen. Curr. Anthropol. **40**, 567–594 (1999)
8. Scott, J.C.: Against the Grain: A Deep History of the Earliest States. Yale University Press (2017)
9. Mokyr, J.: The years of miracles: the industrial revolution 1750–1830. In: The Lever of Riches: Technological Creativity and Economic Progress (1992). https://doi.org/10.1093/acprof:oso/9780195074772.001.0001
10. Mokyr, J.: The later nineteenth century. In: The Lever of Riches: Technological Creativity and Economic Progress, pp. 370–386. Oxford University Press (1992). https://doi.org/10.4324/9781315562315-22
11. Frey, C.B.: The Technology Trap: Capital, Labor, and Power in the Age of Automation. Princeton University Press (2019)
12. Grayson, L.: A brief history of engineering education in the United States. IEEE Trans. Aerosp. Electron. Syst. **AES-16**, 373–392 (1980)
13. Summers, M.: Nicolas Appert. Downs Way Publishing (2015)
14. Parker, M.: Appert – Charles; Francois or Nicolas? Food Technol. **8**, 584–585 (1954)
15. Szczesniak, A.S.: The Nicholas Appert medalists. A reflection of the growth of food science and technology. Food Technol., 144–151 (1992)
16. Corcos, A.: A note on the early life of Nicolas Appert. Food Technol. **29**, 114 (1975)
17. Graham, J.C.: The French connection in the early history of canning. J. R. Soc. Med. **74**, 374–381 (1981)
18. Farrer, K.T.H.: An early Australian recognition of the rationale of canning. Food Aust. **56**, 518–519 (2004)
19. Farrer, K.T.H.: Heat processing 1800–1900: mixed perceptions. Food Aust. **60**, 319–322 (2008)
20. Garcia, R., Adrian, J.: Nicolas Appert: inventor and manufacturer. Food Rev. Int. **25**, 115–125 (2009)
21. Featherstone, S.: A review of development in and challenges of thermal processing over the past 200 years – a tribute to Nicolas Appert. Food Res. Int. **47**, 156–160 (2012)
22. Thorne, S.: The History of Food Preservation. Pantheon Publishing (1986)
23. Cowell, N.D.: An Investigation of Early Methods of Food Preservation by Heat. University of Reading (1994)
24. Cowell, N.: Who invented the tin can? – a new candidate. Food Technol. **49**, 61–64 (1995)
25. Cowell, N.D.: More light on the dawn of canning. Food Technol. **61**, 40–45 (2007)
26. Goldblith, S.A., Joslyn, M.A.R., Nickerson, J.: An Introduction to Thermal Processing of Foods. AVI Publishing Inc. (1961)

Chapter 2
The Invention of Canned Food

Abstract Paris at the turn of the nineteenth century. An entrepreneurial chef perfects a process for preserving food by sealing it in a glass jar with a cork then cooking thoroughly. He struggles in business but writes a book so his methods can be copied. Invention as an evolutionary process. Inventions are only important if they are copied.

2.1 The Inventor[1]

Nicolas Appert was born on Friday November 17, 1749, in Châlons-en-Champagne, a small town about a hundred miles east of Paris, where his parents were innkeepers. Their inn, Le Cheval Blanc, had been in his mother's family for generations but passed to her husband when she married in 1739. In the decade since their marriage, Marie-Nicole Huet Appert gave birth to eight children before Nicolas, but only three survived. After Nicolas, she had two more children—both of whom, mercifully, lived.

Despite these personal tragedies, the family was doing well. Within a few months of Nicolas' birth, his parents moved to a much larger and grander property, L'Hôtel du Palais Royal, in the center of town. It was there Nicolas grew up. His uncle, a priest, taught him to read and write and do some basic arithmetic, and he may have briefly attended school around the age of 10. Everything else he knew he learned from working—first in the family business, then at the even grander La Pomme d'Or hotel nearby, and later in the brewery he briefly ran with his brothers. Most of what he learned there were the practical skills of preparing, preserving, and selling food for a profit.

La Pomme d'Or attracted prosperous and influential guests traveling to and from Paris. King Gustav III of Sweden stayed, as did Stanislaw, the former King of Poland, who Appert taught to make onion soup.[2] It is possible that Stanislaw

[1]The story of Appert's life is taken largely from Malcolm Summers' biography [13].

[2]As an old man, Appert's recalled with pride the surprisingly bland recipe: "Remove the top crust of a loaf, break it into pieces and heat it by a fire on both sides. When the crusts are hot, rub them with a little butter, and heat them again until they are a little roasted; put them on a plate while

J. Coupland, *Containing Nature*, https://doi.org/10.1007/978-3-032-11882-0_2

introduced Appert to his next employer, Count Christian IV of Zweibrücken. At the age of 23, Appert moved to Christian's palace in the Rhineland to work in the kitchen. Although the count died in a hunting accident only 3 years later, Appert continued to serve his widow for another 9 years, both in Forbach on the French-German border and at her fashionable residence in Paris.

Throughout this first period of his career, Appert made his living by feeding wealthy people. He described his education in this way: "Reared in the art of preparing and preserving by these processes, I knew alimentary products; having lived, as it were, in pantries, in breweries, in storerooms, and in the cellars of Champagne, as well as in the factories of the confectioners and distillers, and in the storehouses of the grocers; accustomed to superintend and to conduct establishments of this kind during forty-five years" [1]. He would draw on this practical knowledge, formed over years of practice, as an entrepreneur and inventor.

In 1784, at the age of 34, Nicolas Appert was ready to strike out on his own. He opened a confectionery shop on Rue des Lombards in the center of Paris, which he modestly called La Renommée.[3] A confectioner sold sorts of luxury foods—fruits preserved in sugar, jams, candies, cookies—as well as the elegant tableware expected for the lavish dinners of the wealthy. Appert had a workshop on the premises where he prepared some of the food he sold in the shop. It is likely there, perhaps around 1790, that he began his research on food preservation, which would eventually lead to his invention of the canning process.

But these were troubled times for France. By 1789, a series of bad harvests and bad policies had pushed the country toward famine and revolution. When the Royalist troops were withdrawn from the capital, Appert, perhaps driven by the need to protect his business, joined the militia. As the revolution advanced, he was among the first to join the National Guard and rose to leadership in his local district, the Lombard section. He donated money and arms to the revolutionary cause, participated in police and military actions, and was even selected to attend the trial and execution of the king.

However, political infighting among the revolutionaries soon followed. In May 1793, Appert was denounced. Before he could be arrested, he left Paris on a sudden "business trip" but was captured the following month in Reims. He was imprisoned under dreadful conditions until the radical revolutionary government fell in July 1794. Upon his release, he was able to leave the dangerous world of politics behind and refocus on his business.

Throughout this period, and perhaps surprisingly given the extreme political and economic circumstances, La Renommée had prospered, selling groceries to local people and shipping foods across France. When, just before his arrest, Appert had

frying the onions in fresh butter; ordinarily one puts in three big onions, about 100 g [sic], diced very small; leave them on the fire until slightly tinted, stirring almost continually; next add the crusts an continue stirring until the onion brown. When it is the right color, take from the saucepan, add boiling water, seasoning and any necessary water, then leave simmering for at least a quarter of an hour before serving"—translated by Summers [13].

[3] *The Famous*—I am assured it sounds better in French.

moved to another part of Paris and applied for new documents, he had described himself as a merchant rather than simply a confectioner. Clearly the nature of his business was changing, and he needed new space.

Soon after his release from prison, Appert and his growing family moved a few miles outside Paris to Ivry-sur-Seine, where he established a larger food production facility and began producing the first effectively preserved foods for sale. In 1802, he purchased a much larger property in Massy, about 20 miles to the south. There, he had fields for growing crops, rooms for packing and processing food, and a staff of about 50 people. This was the world's first true cannery.

Appert described the facility as follows: "My laboratory consists of four compartments or workshops. The first furnished with a battery of kitchen utensils, of stoves, and of the necessary apparatus for preparing all the animal substances intended to be preserved... The second is intended for preparing milk, cream, and whey. The third for corking, wiring, and placing in sacking the bottles and other vessels. The fourth is furnished with three large copper kettles mounted in masonry on the furnaces" [1].

By 1807, he is reported as producing a remarkable 20,000 bottles of peas a year, along with other products including beans, fruit, soup, milk, and meat. He was now fully committed to the business of making canned food—and his large factory needed a market to match.

Appert sold his new preserved products in his shop in the fashionable center of Paris, where they were praised by the food writer Grimod de La Reynière: "It is not a question here of describing the process that is used to preserve them in this way. We are only talking about the result, and this result is to have in each bottle and at little cost, an excellent dessert, which reminds us of the month of May in the heart of winter, and often even to be mistaken when it has been made by a skilled cook: it is not an exaggeration to say that peas especially, prepared in this way, are as green, as tender, finally as tasty as those that we eat in the middle of the season" [2].

Appert also used his connections as a merchant to ship his products around Europe, and there are records of him selling preserved food in Russia and Bavaria as early as 1803.

However, Appert's new preserved foods were not cheap. Those processed vegetables he shipped to Russia in 1803 cost about 3.5 francs a bottle at a time when a French laborer might earn less than half a franc an hour [3]. Coarse rye or wheat bread was the staple diet of the French poor in the eighteenth century, and purchasing bread alone consumed approximately half of their income [4]. This already staggering figure—bear in mind the total food budget for an average modern American is only about 10% of income—had risen to almost 90% when bad harvests in 1788 and 1789 drove the country to revolution. These were not people that could buy the new canned foods. The first consumers of canned food were the wealthy few, and Appert needed to find other, larger markets to grow his business.

People who might pay a premium for preserved foods would need to especially value the fact they were preserved. The obvious market was the navy. Fresh food spoiled quickly, and existing methods of food preservation meant the diet on long voyages was poor and frequently resulted in outbreaks of scurvy. Appert sent samples of preserved foods to the officer in charge of naval operations in the Atlantic, who kept them on a ship for 3 months before tasting them and finding them good ("The beans and small peas, prepared with and without meat, have the freshness and agreeable flavor of freshly picked vegetables"). While the testimonial was useful, there were no orders forthcoming—perhaps because the British naval blockade in force at the time meant the French navy was not able to attempt long voyages.

Appert was making products people liked, but the business and political climate in France remained difficult. An economic crisis and mounting debts to friends and family forced Appert into bankruptcy in February 1806. One historian argues that while he continued to operate, he remained in debt for the rest of his life [5]. He sold off some of his valuable equipment and worked desperately to keep his business afloat.

He exhibited his bottles at a trade fair, Exposition des Produits de l'Industrie Française. Although he didn't win a prize, he attracted some useful publicity and press coverage. During the quiet winter season, he toured France desperately trying to generate business. Results were mixed—the navy still weren't interested, but he was able to build a network of retailers willing to buy some of his factory's huge production.

His next step to promote his products would have more lasting consequences. In 1809, he brought examples of his work to the Société d'Encouragement pour l'Industrie Nationale (SEIN, Society for the Development of National Industry) for evaluation. SEIN had been organized in 1801 by a group of scientists, engineers, government officials, bankers, and entrepreneurs with the goal of improving society by supporting technological innovation. Appert hoped his bottled food would be the kind of thing they might support [6, 7].

Without revealing the details of his processes, Appert allowed experts from SEIN to visit his factory and examine his preserved foods. They produced a glowing report, describing the qualities of various products in detail and emphasizing how much better they were than any other preserved foods available. This endorsement, coming from some of the most respected figures in France, brought valuable publicity—but Appert urgently needed financial support.

In May, he wrote directly to the minister of the interior, the Comte de Montalivet, asking for assistance. After another round of investigations by the leading scientists of his day, Montalivet offered Appert a choice: either patent his discovery or publish his methods in a book for the general benefit of French industry, in exchange for a 12,000-franc reward.[4]

Appert took the second path and within a couple of months wrote *The Art of Preserving All Kinds of Animal and Vegetable Substances for Several Years* (also

[4] Many accounts of Appert's invention describe the 12,000 francs as a prize offered by the Emperor Napoleon to solve the problems of feeding his armies, and some even suggest that Appert received the award from the Emperor in person [2, 14–16]. However, Appert's biographer Malcolm Summers, along with other historians, finds no evidence of this having happened. Historian Norman Cowell traces the origin of the prize myth to an 1839 pamphlet by another canner, Fredrick Gamble [17].

known as *The Book for All Households*) in which he revealed his methods to the world: "[F]irst, to enclose in the bottle or jar the substances that one wishes to preserve; second, to cork these different vessels with the greatest care because success depends chiefly on the closing; third, to submit these substances thus enclosed to the action of boiling water in a water-bath for more or less time according to their nature and in the manner that I shall indicate for each kind of food; fourth, to remove the bottles from the water-bath at the time prescribed" [1].

He went on to describe recipes for preparing different types of food for the bottles and how they should be processed. For example, peas: "[t]hey are shelled immediately on gathering. The largest ones are separated from these, after which they are carefully heaped in the bottles..., so as to get in as many as possible. are closed, etc., so as to put them in the water-bath in order to boil for an hour and a half... The large ones which have been separated from the smaller, are likewise put in bottles; they are closed, etc., so as to give them, according to the season, two hours or two and a half hours boiling the water-bath."

To build credibility, Appert included evaluations from SEIN-appointed committees and testimonials from naval officers attesting to the value of his preserved foods. With this publication, Appert made public the foundational process behind all modern canned foods. He was 59 years old.[5]

2.2 Invention as Evolution[6]

How can we best understand what Appert achieved? The two metaphors commonly used to describe technological innovation are evolution and revolution [8]. The evolutionary metaphor draws from biology, emphasizing gradual change in response to environmental pressures. The revolutionary metaphor, by contrast, likens innovation to sudden, radical political upheaval. Appert, who had been a revolutionary in French politics, seemed to view his work as an inventor in similar terms. In 1810, he

[5]The publication of *The Book for All Households* marked a moment of triumph for Nicolas Appert, but world events were soon to bring disaster. Four years later, Napoleon's military defeat led to the occupation of France by foreign forces. In March 1814, Prussian troops seized Appert's factory in Massy for use as a field hospital, and everything—crops, equipment, and materials—was lost. After Napoleon's exile to Elba, Appert returned to Massy and attempted to restart his business. But by March 1815, Napoleon had returned, only to be defeated for the final time at Waterloo in June. This time, it was English troops who occupied the factory—and once again it was destroyed. Appert sold the property and moved his operations to a government-supported facility in Paris. In 1836, he passed the company to a family member and retired to Massy on a government pension. He died there in 1841 at the age of 91 and was buried in a common grave. The company he founded continued to operate until 1964, when it was acquired by the Underwood Company of Boston. The story of that company also plays a role in Chap. 8.

[6]I draw many of the technical details of the history of food preservation related to this question from Chaps. 2, 3, and 4 of Norman Cowell's PhD dissertation [18] as well as Stuart Thorne's book [19].

claimed that his discovery marked a complete break from the past: "…as far as my knowledge extends, no author, either ancient or modern, has pointed out, or even led to the suspicion, of the principle which is the basis of the method I propose" [1].

As we shall see, Appert was mistaken, but we can perhaps forgive his hyperbole. He was struggling to keep his business afloat and needed to be recognized unambiguously as the inventor to get his money from the government. Later food scientists often described Appert in similar revolutionary terms. Katherine Bitting, for example, called him "the creator of a great industry which should bear his name" [9]. Once again, we might forgive her hyperbole. Bitting was trying to tell the origin story of her industry, and the revolutionary metaphor gives a natural way to focus on the "great individual" whose efforts changed the world. In a similar way, the complexities of eighteenth-century politics that led to the American Revolution are sometimes reduced to heroic stories of George Washington crossing the Potomac or chopping down a cherry tree. The simple stories can be powerful as propaganda but miss the realities of how invention—or political change—actually happens.

The evolutionary metaphor, by contrast, emphasizes progressive modification over sudden "eureka moments" of creation. In this view, just as all living organisms evolve from other living organisms through processes of natural selection, all made things evolve from other made things through processes of artificial selection. The inventor looks at the objects available in their world and modifies them to solve a problem. Most modifications fail, but occasionally one succeeds and is adopted—until a later inventor makes a further improvement.

In some cases, this is obviously correct: this year's car, for example, is essentially a small modification of last year's model. Other cases, such the very first motor car or Appert's bottles, are more challenging to understand as evolution. If these are really formed by an evolutionary process, then what are their "technological ancestors" and what, precisely, were the modifications made? Let's look at some of the objects available in Appert's world that he could have drawn on to modify.

Most foods spoil because microorganisms—yeasts, molds, or bacteria—grow in them and cause unwelcome changes in texture, taste, and appearance. Cooking can kill these microorganisms, but when the food cools, it is quickly recontaminated and once again begins to spoil. Traditional means of food preservation—drying, salting, smoking, pickling, and fermentation—change the food so that spoilage microorganisms cannot grow so easily. At the same time, these processes transform the food into something largely unrecognizable as the fresh produce. Many of these foods transformed by preservation have become essential parts of regional cuisines, for example, mushy peas (dried peas), bachalau (dried, salted cod), kimchi (salted and pickled vegetables), or pastrami (salted, smoked beef).

Preserved foods are often packed in jars and bottles for storage and transportation. For example, the Romans transported garum, a fermented and very salty fish sauce, around their empire in ceramic amphorae [10]. The fish was preserved, albeit in a radically altered form, by the salt and by the products of fermentation, while the amphorae provided vessels for transportation and storage. Other preserved foods of this type that would have been familiar in the confectioner's shops of Appert's time include jars of jam (fruit preserved with sugar) and bottles of pickles (vegetables preserved with acid and/or salt).

In all these traditional foods, bottles and jars serve merely as storage containers—they are not central to the preservation process itself. Appert's canning, on the other hand, depended on a combination of technologies to achieve superior preservation: heating (to destroy the spoilage microorganisms in the food) and bottling (to prevent subsequent recontamination of the food). The food is preserved in a cooked but otherwise unchanged form, and the sealed impermeable package is an essential part of the preservation and not simply there to allow storage. But was canning the first way food was preserved by this combination of principles?

One combination of heating and sealing available in Appert's Paris was confits. Meat, most famously duck, is cooked slowly in its own fat and then packed in jars where it keeps for several months. The heat of cooking destroys many of the microorganisms that might otherwise cause spoilage, while the layer of fat on the surface prevents its recontamination by airborne microorganisms. An even more closely related product, cooked sardines packed into tinplate cans under a layer of oil or butter, was sold in France in Appert's time [11]. However, both these methods only offer partial preservation as the foods were highly susceptible to microbial recontamination if the fragile fatty layer was breached.

A second possible precursor of canned foods is bottled fruit. Fruits are easier to preserve than meat because their natural acidity helps inhibit the growth of spoilage organisms. It had long been known that acidic fruits could remain stable when packed in bottles with minimal heat processing. For example, Katherine Bitting quotes a recipe for preserved gooseberries from 1680 that combines heating and bottles in a manner strikingly similar to Appert's: "Gather Gooseberries at their full Growth, but not ripe, Top and Tail them, and put them into Glass Bottles, put Corks on them but not too close, then set them on a gentle Fire, in a Kettle of cold Water up to the Neck, but wet not the Cork, let them stand till they turn White, or begin to Crack, and set them till cold, then beat in the Corks hard, and Pitch them over and they will keep without scalding" [9].

The food scientist turned food historian Norman Cowell tracks versions of this method through English recipe books published in the seventeenth and eighteenth centuries showing some interesting revisions including hammering in the cork before heating and tentative efforts to extend the methods to less acidic, and therefore less easily preserved, foods such as legumes.

These two food preservation methods, confits and bottled fruits, were available in Appert's world and could serve as the immediate "technological ancestors" for his invention of canning—but only if we can be sure he was aware of them. At the time as Appert was working, Thomas Saddington published a method in England for preserving fruit using wide-necked beer or wine bottles which he said were cheaper and more widely available than "what are generally called gooseberry bottles." If such bottles were commonly recognized as being for fruit preservation, that suggests widespread use—at least in England.

However, when Norman Cowell surveyed eighteenth-century French cookbooks, he found several references to preserving meat under fat, but none describing the long-established English methods for preserving acidic fruits. It's entirely possible that Appert was unaware of this prior work. Against this, the food writer Grimod de La Reynière could draw a connection between the inferior existing bottled fruit and vegetables available in France ("already fermented, lost their freshness and their quality during transportation") and Appert's related but vastly improved new products [2]. While Appert was French, he was involved in the trade of luxury preserved foods across Europe and might have had examples in his workshops he could copy and improve. On balance, it seems likely that he was aware of related precursor methods. If these bottled fruits were the immediate "ancestor" of canned foods, what were Appert's critical modifications?

The details of how Appert went about modifying existing bottled foods are not available. Was he selling bottled fruit and noticed that many fermented and tasted unpleasant? Did he try to use an existing recipe for preserving redcurrants to preserve peas only to find that what worked for some foods failed for others? How did he go about changing the conditions, recognizing problems, and trying again? We can only speculate. Doubtless he failed more than he succeeded but eventually came up with a process that was reliable enough to start producing on a large scale and be confident his products would be marketable. Whatever his way of getting there, in his 1810 book, Appert gives a clear statement of what he thought were his two critical innovations—sealing and heating.

His first concern was in getting a reliable seal around the top of the bottle. To do this, he adapted the technology of corking from the wine industry but insisted on much greater care. He recommended using only the highest quality corks, which were to be compressed using a specially designed tool to make them thinner and longer. Once forced into the neck of the bottle, they would expand again, reinforcing the seal.

The bottles were similar to champagne bottles, but Appert specified the glass should be thick to resist cracking and the neck should have an internal lip at the top to hold the cork in place. He designed a special stool with a vice to hold the bottle in place as the cork was beaten into the opening with a wooden bat. For solid foods, which required wider-necked bottles, he created custom seals by gluing together sections of cork. To reinforce the seal, he covered the bottle neck with a specially formulated clay—similar to the wax used on modern wine bottles—or secured the cork with wire, much like the closures used for sparkling wines.

Sealing was so critical to Appert that the only illustrations in his 1810 book showed the tools he designed and used for the purpose. He understood the critical importance of this step from years of hand-on experience which he described in meticulous detail, always writing in the first person.

Appert's second major concern was the cooking conditions. He cooked his foods in a large copper kettle which he kept covered to minimize evaporation. The bottles were packed in sacks, to catch the glass if any broke, and then placed upright in cold water up to their necks. The kettle was brought to boiling, and, after the specified time, the water was drained using a special stopcock at the bottom to prevent overcooking.

He specified the cooking conditions for different types of food. Meat, for example, was first partly cooled and then packed into bottles and boiled for half an hour; peas were boiled for between 1 and 2 hours depending on season; and redcurrants were merely brought to the boil before cooling. He recognized there was a trade-off between effective preservation (cook for long enough) and loss of quality due to overcooking (but not too long).

These two steps, sealing and cooking, were Appert's core technical inventions. Together, they transformed the existing methods of bottling foods, enabling much longer shelf life and making preservation possible for a far wider range of products. Both techniques were borrowed from existing food preparation practices, but Appert's contribution lies in combining them for a novel purpose and specifying, in detail, how they should be applied. This should not diminish Appert's reputation as it is typical for all inventions. Every made thing is made from other made things.

If we accept the evolutionary metaphor for technological innovation, it offers a different way to appreciate the significance of Appert's achievement. In biological evolution, "success" occurs when a mutation provides an advantage and is passed on to future generations in greater numbers. In technological evolution, success is measured by adoption—when others use, replicate, and build upon an innovation.

Appert deserves Katherine Bitting's praise as "the creator of a great industry" not because he invented something completely new but because others copied and improved on his work. Appert had no patent, and his processes were clearly laid out in his June 1810 *Book for All Households*. By late summer, a copy had been smuggled to England; in October, a German edition was published and, in 1811, a Swedish edition. Second and third French editions followed in 1811 and 1813. As Minister Montalivet had hoped when he made publication a condition for funding, Appert's ideas spread rapidly—though Montalivet might have been disappointed that the next major developments occurred outside France.

We recognize Appert as the inventor of canning not because he was the only one working on food preservation but because he serves as a "common ancestor" to all subsequent canning operations. His work became the foundation from which the modern food preservation industry evolved.

Before we leave the topic, it's worth reflecting on the moral dimensions sometimes associated with evolutionary processes and the idea of "progress." Biological evolution is sometimes said to give rise to "fitter" offspring destined to dominate and replace their weaker cousins. This view suggests that evolution drives progress toward "higher" forms of life—and, in the minds of early twentieth-century scientific racists, even higher races of humans. Such thinking has been used as a pseudoscientific justification for the supposed naturalness of social hierarchies. If evolution leads to superior offspring that dominate within a few generations, then the dominance of one group over another might appear to be the inevitable result of inherent superiority.

In reality, "fitter" offspring are simply those mutations better able to survive and reproduce in a given environment. A more virulent pathogen or a tastier variety of strawberry might thrive under conditions of natural or artificial selection, but whether these outcomes are seen as progress or disaster depends entirely on their impact on people. A similar moral undertone often accompanies ideas of technological progress, where new technologies are too readily accepted as inherently better.

The contrast between evolutionary and revolutionary metaphors for technological innovation which I have used here was set out by the historian of science, George Basalla. One of his examples was the invention of the cotton gin by Eli Whiney in Georgia, around the same time Appert was working on canning in Paris. While Whitney's gin is often celebrated as revolutionary, Basalla argues it was merely a modification of existing gins—a slotted plate that enabled faster processing of the shorter cotton fibers grown locally. Unlike Appert, Whitney had a patent, but he was unable to effectively defend it, and others started making new versions of his invention. It was the widespread adoption copies of his innovation that transformed the cotton industry—a textbook example of evolutionary technological progress.

But this example should give us pause to reflect on the positive moral tone associated with the term "progress." Whitney's gin allowed cotton fibers to be separated from seeds much more quickly. That was certainly a boon for the Southern capitalists who grew and traded cotton and for the Manchester industrialists whose factories turned that cotton into cloth. It also benefited consumers, who could now buy better clothing at lower prices. Whether the lives of textile workers improved is debatable, but the massive expansion of cotton production in the American South unquestionably made life far worse for the enslaved Africans who performed the labor.

A technological innovation is successful if it is rapidly adopted, but whether that counts as "progress" depends on what it does and may be different for different people. Katherine Bitting saw modern food processing as a benefit to humanity and Appert as a founding hero. Modern critics of processed foods might look at his innovations as the first step down the wrong path.

2.3 Reflections for Modern Technologists: Entrepreneurship

So, what can food scientists today learn from an entrepreneur in Revolutionary France? Does Nicolas Appert deserve his status today as hero of the profession?

Appert was no scientist. As we have seen, almost all his education was through practical experience working with food. While his descriptions are rich in practical details on preservation methods, they lack scientific method or underlying theory. His work was never concerned with any abstract pursuit of truth, but rather with practical problem-solving and his own business interests. And yet, aspects of what he did offer a much more useful model for modern food technologists than the stereotype of the scientist, aloof in a white coat.

First, Appert was closely involved in making food throughout his career. From working in a commercial kitchen, to a brewery, to a confectionery shop, and to his first cannery, he learned from his close personal involvement with every stage of the process. A modern technologist has more background scientific knowledge to draw from than Appert did but shouldn't lose track of the details of production. I recall visiting factories with a technical manager or laboratories with a chief scientist and seeing them confidently reach into equipment to clear a jam or look at a sample and identify problems before any formal measurements were made. They are successful because they understand the success of any grand strategy depends on the details of how things get done.

Second, while we might praise a scientist today for being objective, Appert had skin in the game. Since opening La Renommée in 1784, Appert had depended on selling products and making a profit to feed his family and pay his staff. He was profoundly interested in making his innovations work. The discoveries we remember him for today had to be developed and managed within the context of a functioning business that needed to deliver products year after year. He had to take risks to innovate and grow, but always with the awareness that failure couldn't be so great as to put him out of business.

Finally, Appert believed so strongly in the value of his idea that he persisted when most people would have given up. He worked for at least 15 years before he received public recognition. He rebuilt his business after two military invasions and one bankruptcy. He spent winters traveling France on terrible roads trying to sell his products to the navy and failing again and again. We leave Appert's story after the publication of *A Book for All Households*, but he continued with his business until he was 87 years old. He quickly adopted metal cans alongside his glass bottles and used salt baths to raise the cooking temperature (see Chap. 3). He was one of the first to experiment with a pressure retort (Chap. 4). Canning was his life's work and, in the words of a later canner, "stick-to-it-iveness-succeeds" [12].

Or at least it does sometimes. When a successful person talks about how they struggled for years before achieving their breakthrough, we can admire their dedication while recognizing their survivor's bias. Most ideas are bad ideas, and the people who spend their lives pursuing them are cranks who rarely get asked for life lessons. While persistence is a virtue a modern technologist could learn from Appert, they should also maintain some perspective and be open to the possibility they are wrong. In the next chapter, we will meet some of the people who copied Appert and recognize that their approach, while less original, offers a better model for modern technologists and entrepreneurs.

References

1. Appert, N., Bitting, K. G. (translator): The Book for All Households; or, The Art of Preserving Animal and Vegetable Substances for Many Years (1810)
2. Graham, J.C.: The French connection in the early history of canning. J. R. Soc. Med. **74**, 374–381 (1981)
3. Wages: Encyclopædia Britannica, pp. 716–729 (1903)
4. Rude, G.: Prices, wages and popular movements in Paris during the French revolution. Econ. Hist. Rev. **6**, 246–267 (1954)
5. Garcia, R., Adrian, J.: Nicolas Appert: inventor and manufacturer. Food Rev. Int. **25**, 115–125 (2009)

6. Butrica, A.J.: Creating a past: the founding of the Société d'Encouragement pour l'Industrie Nationale yesterday and today. Public Hist. **20**, 21–42 (1998)
7. Bishop, P.W.: Who introduced the tin can? Food Technol. **32**, 60–65 (1978)
8. Basalla, G.: The Evolution of Technology. Cambridge University Press (1989)
9. Bitting, K.G.: A benefactor of humanity. In: Bitting, A.W. (ed.) Reproduced in: Appertizing or the Art of Canning; Its History and Development, pp. 7–18. The Trade Pressroom (1937)
10. Pitcher, T.J., Lam, M.E.: Fish commoditization and the historical origins of catching fish for profit. Marit. Stud. **14**, 2 (2015)
11. Brioist, P., La Fichou, J.-C.: sardine à l'huile ou le premier aliment industriel Nicolas Appert et Joseph Colin: une filiation douteuse. Ann. Bretagne Payes. Ouest. **119**, 69–81 (2012)
12. Furman, F.F.: Stick-to-It-Iveness Succeeds. AMG Publishers (1994)
13. Summers, M.: Nicolas Appert. Downs Way Publishing (2015)
14. Long, A.: Chef founds canning industry. Sci. News Lett. **62**, 250–251 (1952)
15. Skolnik, H.: History, evolution, and status of agriculture and food science and technology. J. Chem. Doc. **8**, 95–98 (1968)
16. Stumbo, C.: Thermobacteriology in Food Processing. Academic Press (1965)
17. Cowell, N.D.: An Investigation of Early Methods of Food Preservation by Heat. University of Reading (1994)
18. Cowell, N.D.: More light on the dawn of canning. Food Technol. **61**, 40–45 (2007)
19. Thorne, S.: The History of Food Preservation. Pantheon Publishing (1986)

Chapter 3
The Invention of the Tin Can

Abstract The tangled path from Paris to London in the early 19th century. Appert's invention is copied and improved. Metal cans become commercially important in the niche market of naval expeditions. Microinventions and macroinventions, sub-technologies and innovations. The phylogenetic tree of canning.

3.1 The Inventors[1]

Nicolas Appert published his *Book for All Households* in June 1810. Just 2 months later, Peter Durand secured an English patent: "First, I place and inclose the said food or articles in bottles or other vessels of glass, pottery, tin or other metals or fit materials; and I do close the aperture of such containing vessels so as compleately to cutt off and exclude all communication with the external air… Secondly, when the vessels have been charged and well closed, I do place them in a boiler…which I gradually heat to boiling and continue the ebbulition [boiling] for certain time which must depend on the nature of the substances included in the vessels…" [1]. It was essentially the same process, with the important modification of packing in metal cans and ceramic jars as well as in Appert's glass bottles.

In fact, Peter Durand did not conduct the work he described. He was the son of French Huguenot immigrants—he was christened Pierre—and was working as the agent for Phillipe de Girard [2], an engineer and an entrepreneur. Girard was already known as the inventor of a machine for weaving flax, and he would go on to develop improvements to steam engines and even an early machine gun. Taking the English patent for canning through Durand was simply more convenient during time of war as Durand was English while Girard was French.

Still, while any successful technology will sooner or later be copied, this was astonishingly fast—especially considering that the two countries were at war and a channel blockade was in place. The question of how this patent happened so quickly

[1] Norman Cowell's detailed description of the early development of English canning in Chap. 7 of his PhD dissertation is used as the basis for much of this section [19].

J. Coupland, *Containing Nature*, https://doi.org/10.1007/978-3-032-11882-0_3

has intrigued historians—was Girard an independent inventor, was he a technology pirate stealing Appert's ideas, or was he working with Appert's blessing to surreptitiously extend his business into enemy territory?

We can quickly discount the idea that Girard was completely unaware of Appert's discovery. The timing is too close, both men were working in France and moving in the same social circles, and the processes described were so similar—even down to the smallest detail. For example, in his 1810 book, Appert put great emphasis on the use of composite corks for wide-mouthed bottles: "stoppers of three or four pieces of cork… placed together the way of the grain, the pores of the cork being placed horizontally" [3]. Compare this to the way the same process is described in the Durand patent: "I make use of corks, formed of pieces glued together, in such a manner as that the pores of that substance shall be in a cross direction with regard to the aperture into which such corks are to be driven." This was no coincidence.

However, the Durand/Girard patent goes further, continuing "…and I do also … make use of stoppers fitted or ground with emery, or screw caps with or without a ring of leather or other soft substance between the faces of closure, and also of corks or cross plugs or covers of leather, cloth, parchment, bladder, and the like." Just as he had done by adding metal cans and ceramic jars to Appert's bottles, Girard appears to have copied Appert's work on corks but also significantly extended it.

We can get more sense of Girard's role as an improver of Appert's invention from comments Durand (once again presumably speaking for Girard) made the following year: "When I received from a friend abroad,[2] more than a year ago, a communication of the discovery above described, I perceived that there was still a great deal to be done to render it perfect, …. I substituted tin-cases instead of glass jars or bottles, and prepared to the extent of thirty pounds of meat at once" [4]. While Appert's skills were in food preparation, Girard already worked in metal and machines, so it makes sense that he would be able to quickly extend some aspects of the discovery.

The second question is whether Girard was operating with Appert's blessing or not. Earlier in 1811, Girard had personally demonstrated his metal cans at the Soho Square home of Sir Joseph Banks, the president of the Royal Society. In his letter of introduction to Banks, Girard attached some of Appert's testimonial letters from his *Book for All Households* but removed all mention of Appert's name from them. Perhaps he was cutting Appert out of the deal? From this and from other inferences, Norman Cowell argues that Girard was acting independently of Appert, although the conclusion must remain somewhat speculative [5].

Clearly, much of the 1811 Durand patent was based on *Book for All Households*, but, given the time needed to do the additional work and complete the patent application process, it seems very unlikely Girard could have simply copied it after its publication. He must have had some prior knowledge. How the information reached Girard is unclear, but Norman Cowell points Claude Berthollet—a leading French chemist of the time and a favorite of Napoleon—as the man making the connection.

[2] The "friend abroad" language is also used in the patent. It was commonly used in patent literature at the time when the applicant wanted to get protection in Britain for an invention already in use overseas.

Berthollet was a supporter of both Girard and Appert and was also a Fellow of the Royal Society in London and a friend of Banks.

Regardless of how Girard came to be involved, the improvements to Appert's work he described were significant. More importantly, he was the one with the patent and was in a position to make money from it. However, instead of going into business himself—perhaps a difficult task for a Frenchman in London—he sought English collaborators. The men he found had prior experience with similar Anglo--French partnerships.

At the start of the nineteenth century, Britain was at war with France and communications between the nations were difficult. One person who could move backward and forward relatively easily was John Gamble (1777/1778–1859). Gamble represented the British Army in Paris in matters of prisoner welfare and repatriation. However, he wasn't above using his position to conduct some personal business on the side. One of his first ventures was smuggling out a description of a continuous papermaking machine recently invented in France.

To establish the papermaking business in England, Gamble turned to the Fourdrinier brothers, wealthy Huguenots, for financing and to John Hall (1765–1836), a paper mill owner and engineer, for technical support. Hall quickly passed the engineering problem on to his former apprentice, business partner, and brother-in-law, Bryan Donkin (1768–1855). In 1803, Donkin opened a factory to make the paper machines in London, while Gamble moved to nearby Hertfordshire to produce paper. Although Gamble's paper business collapsed in 1811, he soon led his partners into a new venture using another French technology exported under questionable circumstances—canning. In 1812, Donkin, Gamble, and Hall bought the rights to Girard's patent (held in Durand's name) for the significant sum of £1000 and used their temporary monopoly to go into business making canned food.

As we saw in the last chapter, canned foods in this period would never be cheap foods. In addition to the costs of the food itself, making the metal can mean hiring a tinsmith, and the long cooking times meant that relatively few could be processed in a day. The new product was therefore unsuitable for mass consumption, but, just as Appert had seen in France, there was an obvious niche market—naval provisioning.

In this context, canned foods enjoyed several important practical benefits over alternative rations available at the time—particularly meat. It was possible to preserve meat for long sea voyages by salting it, but salted meat required a lot of fresh water to prepare. Portable (dried) soups were sometimes used, but were not popular—sailors sometimes called them "veal glue" [6]. Instead, many ships carried live animals aboard to be butchered when needed, but they were difficult to care for and required their own fresh water and fodder—all taking up space in a crowded vessel. Live animals were also time-consuming to process, and their skin, bones, and intestines were waste. Canned foods on the other hand could simply be opened, heated, and eaten.

Canned foods had the additional advantage in that they could be shipped and stockpiled around the world—freeing captains from dependence on local food supplies in remote ports which might be scarce, expensive, or entirely unavailable. Cowell quotes Commodore Evans writing from Bermuda in 1814: "On calculation, we think it is cheaper at 2/6 per lb than any meat we can get here. We pay … about

20d sterling for all we get; more than half is bone, a good deal spoils on hand; and if we kill our own, what with feeding, sickness, deaths and losses of one kind or another, we think Donkin's infinitely cheaper."

Before they could earn Navy contracts, the London canners needed to attract the attention of the British establishment. They employed two related strategies: securing the support of influential individuals and, if possible, demonstrating the value of their products on real voyages. Girard's initial approach to Joseph Banks and the prominent members of the Royal Society in Soho Square in early 1811 was part of this plan. The Royal Society, founded in London in 1660 to advance science, had been led by Banks since 1778. Banks had gained fame on James Cook's expedition to the South Seas, where he was among the first Europeans to visit Australia. He was wealthy, well-connected, and highly influential. If he said something was important, people listened.

In November 1812, a sea trial was organized, with cans sent on a round trip to Jamaica. Upon their return to London, the few available samples were used strategically to build credibility. One can was sent to the Duke of Kent, the son of King George III, and shared with other members of the royal family. A second can was sent to the Victualling Board of the Lords Commissioners of the Admiralty, the officials responsible for purchasing supplies for the Navy. Both were impressed, and the Victualling Board requested Donkin, Hall, and Gamble to provide more samples for larger trials with multiple Naval squadrons stationed around the world. The Royal Navy was ready to buy at least some canned food, and Durand's patent meant only one firm was able to provide it.

3.2 The Preservatory

In 1813, Donkin, Gamble, and Hall built their canning factory—sometimes called a "preservatory"—on the Bermondsey site in South London which Donkin had previously used for his papermaking machines. We can get a sense of what it looked like from the report of a visitor: "Their kitchens for cooking, and larder for cooling the meat before it is packed up, are very complete, and they have considerably improved upon the French mode above-described, by substituting, in most cases, canisters of tinned iron for the glass jars and bottles; they are made by very carefully soldering up the tin plates to form circular or oval boxes, of dimensions suited to what they are intended to contain; and these, when filled with the prepared meat, have lids soldered down upon them very closely: they are then boiled in the water-bath, but do not require that precaution of gradually heating and gradually cooling, which must be observed with glass vessels, which would otherwise be subject to break. The canisters may be put into the bath when the water is boiling, and taken out when finished, another set being put in, so as to keep constantly going. Immediately the canisters are withdrawn, they are carefully examined, to discover and stop the leaks, and the boxes are then well varnished, to prevent rust" [7]. The use of "tinned iron" was their key innovation.[3]

While iron is strong, easily worked, and impermeable, it is prone to corrosion from exposure to the acids in food. Tin, on the other hand, is resistant to corrosion but expensive and too soft to form a can by itself. A composite material, tinplate, is needed.

Cornish tin had been valued since antiquity, and the process of tin plating had been commercialized in Britain in the seventeenth century [8]. By the early nineteenth century, South Wales had become a center for the manufacture of high-quality, affordable tinplate [9]. It was made by rolling heated iron bars into sheets, hammering them to the desired thickness, dipping them in molten tin to achieve a very thin coating, and then carefully cleaning them before sale [10]. The tinplate used to make a can in 1824 was 18.5 thousands of an inch thick, but its tin coating was only 0.5 thousandths [11]—these figures would also be typical in modern cans.

The first cans were made by hand at the preservatory.[4] A rectangle of metal was shaped for the can body using shears and then bent around a wooden form to make a short cylinder. Disks of metal were cut for the ends using circular shears, and a flange was added by beating it into a mold. Wooden mallets were preferred over metal ones to bend the tin without scratching it. A filling hole about half an inch in diameter was cut in one end of the can using a punch, and a cap was made to cover it. The edges of the body were overlapped, clamped, and sealed closed with a lead/tin solder to make a side seam. The top and bottom ends of the can were pushed onto the body and soldered in place to make a seal. Much of this work of can-making could be completed before the food was ready to be packed, and it was conducted either by factory workers or specialized tinsmiths in the off-season.

When it came time to pack the food, the can was filled by pushing solid pieces—chunks of partially cooked meat and vegetables—through the filling hole in the end. Hot sauce or juice was added to cover the solids, and the cap was sealed on. This was the only soldering task required during actual food production. The method, known as a "hole and cap" can, had a notable drawback: unless the can was especially sturdy and heavily soldered, it would often swell or even burst as the contents expanded during cooking. An related method, called a "hole in cap" can, left a tiny vent hole in the end cap to allow the steam to escape during heating. This prevented the can from becoming pressurized and reduced stress on the seams. The vent hole was quickly closed with a drop of solder, either during or immediately after the cooking operation, to complete the seal. Cooking was initially done using Appert's

---

[3] The visitors apparently missed the significance of the venting process, where a pin hole in the end was left open during the first part of the cook and then closed with solder before the cooking was completed.

[4] The description of the early craft manufacture of tin cans is taken from Chap. 3 of Gregg Pearson's PhD dissertation [10]. While Pearson's description is based on American practices from a few years later, he suggests there hadn't been any substantial changes since the period we are discussing. The processes of filling and cooking the cans are from the historic documents reviewed by Cowell [19].

boiling water kettles, though salts were soon added to the water to raise the temperature (described in more detail in Chap. 4).

While Donkin, Gamble, and Hall experimented with many types of foods, their main interest was canned meat for naval rations, and they produced a range of canned beef, lamb, and soups. They couldn't produce cans in the volume or at the price required to feed all of the sailors, so their initial products were special rations for the sick and injured or for polar expeditions. Contracts from the Admiralty varied dramatically from only £147 in 1816 to £9099 in 1821 (to supply an attempt at the Northwest passage). In this period, the company made about 12,000 large—between 4 and 20 pound—cans per year, 40 or 50 a day. It wasn't much, but they were in business selling canned food.

We will leave the story of the London preservatory in 1824, the year the Durand patent expired. By that time, Peter Durand had died, and Bryan Donkin had left the business which continued trading as John Hall and Company. Donkin went on to achieve great success as an engineer, contributing to many of the important infrastructure projects of early Victorian Britain [12]. He was elected a Fellow of the Royal Society in 1836, and his company continues today as Bryan Donkin Valves and as part of other engineering firms. The canning business itself was eventually acquired by what would become Crosse and Blackwell.

We are lucky to have tangible evidence of what John Gamble and Company was doing in 1824. British polar expeditions had benefited from the introduction of canned foods, which allowed ships to overwinter in the Arctic and so complete longer voyages. The explorer William Parry had been so impressed by Gamble's products on a previous voyage that he specifically lobbied to use them again on his 1824 expedition. Gamble duly prepared and delivered the food needed—including a 4-pound can of roast veal.

The can traveled to the Arctic onboard HMS Hecla but was returned unopened and left in storage for years, eventually ending up at the Museum of the Royal United Services Institution. There it remained until 1939, when Jack Drummond, a professor of biochemistry at University College London, was granted permission to open it [13]. The original label was still present, with instructions to the cook to "cut round on the top near the outer edge with a chisel and hammer." Drummond's team instead used a specially built can opener.

The meat inside appeared to be in very good condition, although the color was a salmon-pink that faded to a more typical cooked veal color on exposure to the air. Most of the fats had hydrolyzed to form free fatty acids, so, even if it were safe, it would have been disgusting to eat [14]. Despite the toxic reputation of the lead-based solders used at the time, the meat only contained 3 ppm lead—less than the current FDA limit for bottled water.[5] Viable bacterial spores were isolated from

[5] Lead from the solder in canned foods has been implicated in poisoning the explorers of this period. Most famously, when the body of John Torrington, a sailor on Franklin's 1845 expedition, was exhumed from Beechey Island in the Canadian Arctic in 1986, it was found to contain high levels of lead. Discarded cans from the expedition were found in the area with solder in direct contact with the food. Furthermore, stable isotope analysis revealed the lead in the solder matched

the can contents suggesting the cooking processes used in 1824 would not meet modern standards [15].

The processes Donkin, Hall, and Gamble had developed at the London preservatory, while far from perfect, produced food that was highly valued at the time and sufficiently preserved to be recognizable and perhaps even edible 115 years later. Food could be harvested in one place and time and consumed as essentially the same thing in another, and people would pay for the privilege. There was a market for canned food which, after 1824, anyone was free to compete in.

3.3 Metal Cans as Technological Innovation

In the previous chapter, we used the example of Appert's achievements to look at what it means to invent a technology. What Appert did was not completely novel, but by combining existing things—bottles, corks, cooking vats—to do something new and it made an impact. He published his result and showed the system worked. In this chapter, we looked at modifications to that technology first by Girard and later by the group of London canners with Donkin as their lead engineer. By comparing these two sets of achievements, we can learn new lessons about technological development.

The economic historian Joel Mokyr makes a distinction between "macroinventions," i.e., dramatic steps forward that appear suddenly and apparently without clear precedent, and "microinventions," i.e., subsequent modifications to existing macroinventions to make them work better [16]. At one level this distinction seems problematic. If our conclusion from the last chapter that all made things are really developed from other made things is correct, then nothing can be invented "without clear precedent."

However, Appert's method illustrates how existing technologies can be repurposed to serve entirely new functions, opening up a world of possibilities. Before Appert, bottles and corks and boiling pans of water could only mean wine and cooking. After Appert, the same technologies could also mean food preservation. The fact that the existing technologies were now doing something new meant that anyone trying to improve them had a different goal and would make different types of modification. For example, Appert's decision to modify his bottles to have wider necks was a technological innovation that only made sense when the purpose of the bottles had been changed from storing wine to preserving solid foods. In this light, Appert's bottled food can be seen as a macroinvention, while the subsequent refinements to his method represent microinventions.

We can get a fuller picture of how microinventions work by dissecting the technology of canning and the ways Girard and later Donkin worked on it. The

---

the lead in the body. Although Torrington died from pneumonia, lead poisoning form the canned rations he ate on the ship may have contributed to weakening his constitution [20].

economist W. Brian Arthur argues that all technologies are made up of sub-technologies [17]. For example, the technology of canning requires two sub-technologies: sealing and cooking. In his work, Appert achieved these using glass bottles and corks for sealing and boiling water baths for cooking. Arthur's idea is recursive, so each sub-technology is itself made of its own sub-technologies. For example, Appert's bottle-sealing method is a sub-technology of canning but also has its own sub-technologies: composite corks, the ridge on the bottle for binding the cork in place with wires, and the stool and mallet used for hammering the corks in. Innovation involves replacing a sub-technology in order to improve the performance of the overall technology. When Girard replaced glass bottles with soldered metal cans, it marked a major innovation—one of the higher-level sub-technologies was completely reimagined using a different material.

The London canners were better placed than Appert to make all this work in a factory setting.[6] As we have seen, they had good Welsh tinplate readily available, and they even had prior experience with the material. Donkin had described a method for tinplate in his notebooks in 1808, and his partner, John Hall, had previously established an iron foundry.

Tinplate was already widely used to make boxes and containers by joining pieces together with solder. The challenge in this new application was that heating a sealed can increases the internal pressure, stressing the seams. Even a tiny, invisible crack could be enough to admit bacteria and cause the process to fail. Resisting internal pressure was a challenge the tinsmiths had never had to deal with before, and overcoming it required technical innovation. Their key microinvention was the vent hole in the end which was left open during the initial stages of cooking and then sealed with a drop of solder. By allowing the hot gases to escape before sealing, the internal pressure did not increase during cooking, and the stresses on the seams were lessened. Because only this tiny hole had to be sealed that the can could be easily closed while it was still hot, preventing any recontamination of the contents.

Once the metal cans became reliable, they offered several advantages over glass. Firstly, glass bottles tended to crack if placed directly into boiling water, so cooking required placing them into cold water and then heating. Metal cans on the other hand could easily be moved in and out of a continuously boiling water bath, saving time and fuel.

A second advantage of metal over glass was although the metal cans were opaque, their slight flexibility gave some useful insights into what was happening inside. When the cans cooled, the contents contracted, and the ends curved inward slightly in response. Readers familiar with home canning in glass jars will recognize the distinctive sound of the metal lids popping inward as the food cools. The London canners learned to recognize concave ends as evidence that there was a good seal. In contrast, while Appert knew that sealing was important, he had no immediate

[6] Appert did start canning in metal later, but only after visiting London in 1814 and seeing the process there. In the 1831 edition of *Book for All Households*, the last published in his lifetime, Appert shows the same meticulous focus on the practical details of working with tinplate that he had earlier applied to canning in glass [21].

way to know if his glass bottles were properly closed. The London canners also began holding their cooked cans in heated warehouses for a period before delivery, watching for signs of swelling caused by bacteria that had survived the cooking process. Glass bottles would require much greater internal pressure before they failed. Both of these quality control processes are still used today to detect failures of seaming or the survival of bacteria after cooking.

Finally, metal cans had specific advantages over glass in their particular market—naval rations. They were both tougher and lighter than glass, making them easier to store on ships. In addition, because metal cans could be opened by cutting off their ends, they could be used for the large pieces of meat the sailors preferred.

However, metal cans aren't unequivocally better than glass bottles in all cases, as they are simply different. Indeed, glass has advantages over metal that remain important in certain applications. Glass allows the preserved food to be seen, which might be especially attractive for the brightly colored fruit sold in Appert's Paris shop or the fateful jars of olives we will meet in Chap. 9. Bottles also don't require skills with solder and tin, making them accessible for someone who just wanted to pack a small amount of produce. Canners continued to pack in glass throughout the nineteenth century, and glass remains the method used by both home canners and for a few specialized commercial products today. While the focus of this book is on the development of canning in metal, that shouldn't imply that canning in glass is either outdated or static. It, too, continued to evolve and improve with its own microinventions, for example, the introduction of screw caps and rubber gaskets, which made sealing more reliable [18].

Perhaps we shouldn't view microinventions like the metal can as mere improvements on Appert's canning method, but rather as the birth of an alternative approach to canned food. The older technology did not disappear; before the London canners, there was just glass—after them, there was a choice between metal and glass. To pick up on the evolutionary metaphor for technological development introduced in the previous chapter, each significant microinvention is analogous to the separation of one species into two, creating a new branch in the phylogenetic tree. Each modified technology may thrive in different economic niches, and each is subject to evolution through its own set of microinventions.

A subsequent branch in the evolutionary history of canned foods was the introduction of the flexible pouch by the US Army during the Vietnam War which made soldier's rations lighter. The tough plastic and metal laminate pouches replaced the sealing sub-technology in Appert's canning macroinvention, and so they form another branch on the same phylogenetic tree. Pouches did not cause the metal cans to "go extinct"; both technologies continue to thrive in different applications. The flexible pouches themselves facilitated a later microinvention, high pressure processing, which replaced the cooking sub-technology in Appert's canning macroinvention. Juices and guacamole sealed in pouches and sterilized with high pressure are simply the latest in a growing family of products that have evolved by artificial selection from Appert's discovery.

In the last chapter, I argued that Nicolas Appert deserved his fame because his book serves as a "common ancestor" for all modern canning. Recognizing his contribution extends to other branches of the phylogenetic tree founded on canning only serves to increase his reputation.

3.4 Reflections for Modern Technologists: Making Things Better

Joel Mokyr argued that macroinventions are very rare and depend so heavily on a mixture of luck and genius that they are essentially unpredictable. With this in mind, we should reconsider the praise offered in the last chapter to Appert for his persistence. Yes, in his case, spending decades of his life engaged on a single problem led to success, but most people showing this sort of dedication will fail and be remembered as cranks or, more likely, not remembered at all. Pointing out the long odds against breakthrough success will not discourage some people from pursuing it—lottery tickets, after all, sell well—but their strategy is not a rational one.

Microinventions, in contrast, are more predictable because they aim toward a clear goal. If you understand how an existing machine works and what it is supposed to do, you can set about improving it. Girard, followed by Donkin with Gamble and Hall, recognized such an opportunity and used a combination of technical knowledge, personal connections, and economic courage to make it work. Their strategy was more rational than Appert's.

Their approach to problem-solving was also very different to Appert's. While Appert apparently worked alone for decades, the London canners worked together toward a common aim. It appears Girard and then Donkin took the lead in the technological innovation work, but the others were important too as connectors, as managers, and as investors. They made progress much more quickly, buying the patent one year and selling products the next. An individual might invent something, but technological innovation more often requires a group.

And while they lack the glory of being the inventors of canning, what the London group achieved was essential. After all, it is only through the accumulation of microinventions that the original macroinvention becomes really useful. Or, as Mokyr more elegantly phrased it: "Without subsequent microinventions, most macroinventions would end up as curiosa in musea or sketchbooks" [16]. When Girard and London canners examined Appert's assembly of technologies, the macroinvention, they saw a potential for profit, but only if certain modifications—microinventions—could be made. As Durand, speaking for Girard, said: "I perceived that there was still a great deal to be done to render it perfect" [4].

Modern technologists make their living by trying to do much the same thing, so perhaps it is these forgotten London canners rather than Appert himself that should be the ones celebrated as heroes of the profession and as models of good technology work.

References

1. Durand, P.: Preserving Animal and Vegetable Food (1810)
2. Cowell, N.: Who invented the tin can? – a new candidate. Food Technol. **49**, 61–64 (1995)
3. Appert, N., Bitting, K. G. (translator): The Book for All Households; or, The Art of Preserving Animal and Vegetable Substances for Many Years (1810)
4. Durand, P.: Observations of the patentee. Belfast Mon. Mag. **7**, 219–220 (1811)
5. Cowell, N.D.: More light on the dawn of canning. Food Technol. **61**, 40–45 (2007)
6. Laing, E.A.M.: The introduction of canned food into the Royal Navy, 1811–52. Mar. Mirror. **50**, 146–153 (1964)
7. Rees, A.: Provisions. The Cyclopedia (1819)
8. Parkinson, C.N.: Tinplate – an outline history. Hist. Today. **7**, 610–617 (1957)
9. Bell, R.A.: Origins of the canning industry. Trans. Newcom. Soc. **38**, 145–151 (1965)
10. Pearson, G.S.: The Democratization of Food: Tin Cans and the Growth of the American Food Processing Industry, 1810–1940. Lehigh University (2016)
11. Lewis, W.R.: Investigation of the metal container. In: Historic Tinned Foods, pp. 35–40. International Tin Research and Development Council (1939)
12. Anonymous: A Brief Account of Bryan Donkin F.R.S. and the Company he Founded. Bryan Donkin Company Ltd. (1953)
13. Drummond, J.C., Lewis, W.R.: Historical introduction. In: Historic Tinned Foods, pp. 1–21. International Tin Research and Development Council (1938)
14. Drummond, J.C., Macara, T.: Chemical investigations. In: Historic Tinned Foods, pp. 22–28. International Tin Research and Development Council (1939)
15. Wilson, G.S., Shipp, H.L.: Bacteriological investigations. In: Historic Tinned Foods, pp. 29–34. International Tin Research and Development Council (1939)
16. Mokyr, J.: Introduction. In: The Lever of Riches: Technological Creativity and Economic Progress, pp. 1–16. Oxford University Press (2002). https://doi.org/10.7312/zhao12754-003
17. Arthur, W.B.: The Nature of Technology: What It Is and How It Evolves. Free Press (2009)
18. Rzepecki, K.: The Mason Jar: Preserving 160 Years of History. masonjars.com
19. Cowell, N.D.: An Investigation of Early Methods of Food Preservation by Heat. University of Reading (1994)
20. Kowal, W., Beattie, O.B., Baadsgaard, H., Krahn, P.M.: Source identification of lead found in tissues of sailors from the Franklin Arctic Expedition of 1845. J. Archaeol. Sci. **18**, 193–203 (1991)
21. Appert, N.: Le livre de tous les ménages. Chez Barrois l'ainé (1831)

Chapter 4
The Industrialization of Tin Cans

Abstract Largely America from 1824 to WWI. The processes of making and cooking metal cans are modified to replace skilled labor with integrated systems of machines. Industrialization. The evolution of technology by a hierarchy of microinventions that encompasses even the smallest technical modification.

4.1 A Century of Microinventions[1]

The metal cans described in the last chapter were the first microinvention used to improve Appert's process. Further microinventions over the following century would transform the London preservatory into modern canning factories. To get a sense of the magnitude of the change, think about the early nineteenth-century process of making a food can from tin-plated iron sheets. A tinsmith would cut out a rectangle of tinplate with shears and roll it around a wooden form. Next, they would overlap the edges and seal the join with solder to make a cylinder. They would cut two discs of tinplate as ends and cut a filling hole in one of them with a stamp. They would bend over the edges of the can ends by hammering the disk into a wooden form to make a shallow lip that could be pushed inside the cylinder and soldered to make the final can. In the early years of canning, this work would be done by the canners themselves, usually in the off-season when there was no product to pack, or by local tinsmiths on a contract basis. It might take a tinsmith 10 min to make a single can.

[1] This chapter is focused on the manufacture of cylindrical three-piece steel cans (a body and two ends) although many other forms are common (e.g., two-piece cans—a body and lid—and the aluminum cans used for beverages).The processes of metal can manufacture in the nineteenth and early twentieth century in America are described in detail in Chapter 3 of Gregg Pearson's doctoral dissertation [18]. Although much of the focus of this chapter is on canning in America, technical advances in England in the same period are described in Chapter 5 of Stuart Thorne's *History of Food Preservation* [8]. For the insights on the operation of a modern can-making factory, I am grateful to Nathan Pearlman for a tour of the Can Corporation of America plant in Reading, Pa.

J. Coupland, *Containing Nature*, https://doi.org/10.1007/978-3-032-11882-0_4

Next, visit a modern can-making factory. Various grades of steel, plated with tin or chromium, are delivered in massive rolls—each as tall as a person and so heavy that a truck can typically only carry two. The coils are carefully opened and cut into sheets for use either as bodies or as ends. A thin layer of lacquer is applied by an industrial coater, and the sheets are baked to protect the metal from corrosion and the food product from spoilage.

Disks of metal are stamped out for ends and shaped to form the "curl" that will later be rolled into the double seam. The inside of the curl is filled with a compound to create a gasket-like seal, and the ends are baked to cure it. Other parts of the lacquered sheets are cut into rectangles that will form the can bodies, which are then rolled and welded into cylinders. The cylinders are then flanged, strengthened by forming ridges (beads) into the metal, and one end is attached with a double-seamer. The seams are tested with compressed air using an in-line tester, and the empty cans are ready to be shipped.

All these tasks are done by different machines, and the parts are moved between machines on magnetic conveyors. The machines are supervised by workers who clear jams and do spot checks on quality. At the center of the factory, a testing laboratory receives samples from different points along the production line and uses sophisticated measuring instruments to ensure everything is within specifications. A modern production line can produce 300–500 cans a minute from the labor of perhaps 50 people.

It wasn't a single innovation that led to this change, but rather a series of small steps, further microinventions, made by people who are now largely forgotten. It is the accumulated effect of these steps that changed canning from Appert's handcrafted bottles to Warhol's mundane cans in the twentieth century.

You can get a flavor of what motivated the changes by imagining yourselves in the role of a London cannery owner in around 1824—the end of Chap. 3. You're trying to compete for navy contracts against other canners making essentially the same products with essentially the same processes. If you can reduce your costs, you can compete on price. One way of doing this is to pay your workers as little as you can get away with while at the same time ensuring they make as many cans as possible in the time you pay them. Perhaps you could incentivize speed in your workforce by making at least some of their wages come from piecework? Or maybe you could invest in new tools that would help your workers produce more quickly. What small changes might give you the edge?

To get a sense of the process you are trying to improve, try it yourself now. Sit down with some glue, some card, and some scissors, and try to make a "can" like a tinsmith in 1824 might have done with tinplate and solder. Your first one is probably a mess, but you get better as you practice and become more skilled. You start to think of ways to organize your work better. Why not make a "benchmark" on your work surface to measure the size of a body without needing a ruler every time? Can you improvise a wooden form to bend the edges of the round end to form a lip for gluing? What if you used a paperclip to hold the edges of the body together while you glue them?

What if you teamed up with a couple of friends and divided the labor—one person cutting rectangular bodies, another cutting round ends, while you focused on gluing? By learning from experience and relying on your own resourcefulness, you could quickly improve your output using better tools and more efficient practices. And there's no reason to slow the pace of improvement.

Perhaps you could invest some money to have someone build a guillotine to cut rectangular bodies without using scissors or a stamp to cut perfect circles for the ends. Maybe they could even attach a power source to these cutters, allowing you to produce hundreds of pieces at once from a stack of card. While the first round of innovation could be made by workers and managers within the factory, these later machines require dedicated designers, engineers, and specialized workshops to build them. They are more expensive and require careful planning, so they're only possible if someone has both the financial resources and the authority to decide. The friends you shared work with have become your employees. But there's still no reason to slow the pace of improvement.

Perhaps someone can invent a machine that can roll the card so a worker can simply apply the glue, or perhaps they could even add a glue applicator to that machine and do away with the worker completely? Your new machines are now making cans so quickly that you have saturated the local market for cardboard cans. You should build a new factory with good access to the transportation networks that can take your products to the people that need them.

Our imaginary sequence of microinventions have shifted production from many skilled workers with simple tools to systems of complex machines in large factories operated by a few workers doing simple repetitive tasks. We have taken a craft process of making cardboard cans and industrialized it. And still someone, somewhere, is looking at how their cans are made and looking for yet more improvements.

In a real cannery, the owner will benefit from each stage of this process of industrialization—but only if they stay ahead of any competitors. If they do, their costs of production per can decrease and their profits increase. If they don't, and their competitors can lower their cost of production more quickly, they risk being undercut on price and going out of business. The people who buy the cans also benefit as competition between can-makers means the cost of their products is driven down. The workers making the cans might benefit, but if they can organize to force the owner to share the value gained by their increased productivity. However, a second group of workers may benefit more from the process of industrialization—the technologists who work to design and maintain the machines and to ensure the new processes work in factories. This second type of skilled technology work is a major focus of this book.

This chapter will show how the processes of industrialization worked themselves out historically in the development of the two "shared" technologies that apply to most forms of canning—making cans and cooking cans.

4.2 Making Cans

Some of the first tools built for can making were adapted from the tools tinsmiths were already using to make other products, for example, foot-operated metal cutters which produced rectangles of tinplate more quickly and precisely than handheld shears. The worker would simply push the tin plate up against a standardized guide marked on the workbench and then press the pedal to cut a perfect rectangle every time.

As canning became a larger industry, people started building cutting machines specially designed to meet the growing demand for precise can parts. For example, in 1847, Allen Taylor built a circular cutter to make disks of metal for can ends. Once these new machines for making can parts were introduced, they themselves underwent incremental improvements through their own evolutionary processes. In 1849, Henry Evans added a pendulum press to a circular cutter that produced enough force to simultaneously add a metal flange for seaming the end onto the body. Later, William Numsen added a second cutter to the machine that simultaneously punched out the filling hole from the end.

Similar simple machines were made to speed up the process of soldering the parts together, starting with forms and clamps to hold the piece in place so the worker could use both hands to apply solder. For example, the Jones block developed in the 1850s held the edges of a can body in place with an overlap ("lapped"), so the worker could easily apply solder to the seam. Pearson estimates that effective use of these bench tools increased worker productivity from 6 to about 60 cans per hour by 1860.

Going faster still required moving can production from an individual workbench to more sophisticated floor-standing equipment, powered by belts from a water wheel or steam engine. By the 1880s, a mechanically powered press for can ends could produce 120 per minute, while a foot-powered model from the same company could only make perhaps 20. Although the powered machines were significantly more expensive ($120 vs. $50), the costs were offset because fewer employees were needed. In another example, an end-soldering machine—the Howe floater—could seal 150 cans per hour and only needed two operators: a skilled mechanic and an apprentice. Indeed, many advertisements from this period emphasized that the machines could be operated by "a smart boy." No matter how smart these boys were, they would not be paid as well as the men they replaced.

Adding a machine could improve one step of the process, but the overall speed of can production in the factory would still be limited by the slowest, least mechanized step. For example, if your plant had a foot-powered end cutter (20 cans per hour) and a Howe floater (150 cans per hour), the overall rate of making cans could be no greater than the pace set by the cutter. Indeed, it might be a good deal slower, as the parts had to be moved around the factory from one machine to the next and frequently required considerable hand work to prepare each part for each machine. For example, a "machine made" can in 1883 required the worker to take a cylinder of metal—the can body—and press one end first into a sponge soaked with hydrochloric acid and then into powdered rosin to form the flux during soldering and finally seat a disk of metal with a lip, the can end, over it [1]. It is only after all these manual steps that the can is ready for the end-soldering machine.

Toward the end of the nineteenth century, can-makers began addressing these bottlenecks by connecting machines into integrated systems of manufacture. One of the first was established in 1883 by the Norton Brothers of Chicago, former tinsmiths who had made cans for the local meat packers [2]. The cutting, forming, and sealing machines they used had all been developed previously and used separately, but the Nortons connected them with conveyor belts to create a continuous production line.[2] The remaining machine operators worked only at the periphery, for example, loading ends onto a conveyor.

When everything was behaving properly, the Norton system could make 3000 cans per hour. The machine made cans were also cheaper, not only because of the reduced labor costs but also because they reduced the amount of solder needed from 18–20 pounds of solder per 1000 handmade cans to less than 6 pounds for machine made [3]. Their quality was also superior, as machines could repeat the same precise movements on standardized metal pieces with consistency. The Nortons even added a seam tester to their continuous line which inverted each can in boiling water, and a worker could look for bubbles as the heat expanded the air trapped inside.

The Nortons did not can foods themselves, but instead focused on making cans and making the machines that made cans. In 1894, they opened a state-of-the-art factory in the Chicago suburb of Maywood capable of making half a million cans per day for fruit and vegetable packing. This huge capacity was only useful because the factory was connected to the rail network, allowing it to easily supply distant canneries.

In the early days of the industry, canners made their own cans by hand, usually in the off-season when their primary food products were not available. By the end of the century, while some canners still made their own cans, a viable alternative had emerged: purchasing premade cans from dedicated manufacturers running their own large, highly efficient plants. The finished cans were essentially the same as the ones used in London at the start of the century, but the way they were made had been evolved from a craft process to an industrial one. The final stage in the development of a modern metal can required a complete change in the ways it was constructed.

Up to this point, all can seams had been sealed with solder, but solder was expensive, slow to apply, and could contaminate the food with lead. Furthermore, filling through the small hole in the end was a slow process, and large pieces of fruit were frequently ripped, leaving the syrup cloudy. Finally, during the process of fixing the

[2] Early observers quickly turned to the analogy of a flowing liquid to describe the movement of cans and can parts around the factory: "Thus there are two streams of cans rolling forward and two coming back below, carried on the cooling belts, the last one discharging the finished cans on a trough, where they roll into the testing machine" [2]. In modern can-making factories, magnetic conveyor belts serve the same function—and it remains disconcerting for a visitor to see a rushing line of cans make an abrupt right angle turn and move up to continue their journey, inverted, across the ceiling.

cap over the fill hole, the heat of the soldering iron often burned the contents leading to unsightly, albeit harmless, black specs in the food. There were lots of reasons to get rid of both solder and the fill hole, but this required abandoning a set of optimized processes to make something new and better.

Tinsmiths had long known how to make a solder-free joint by bending out the body slightly to form a flange at the top and bending the end into a broad hook called a curl. The end curl was rolled around the body hook and then pressed together to form a "double seam." However, the joint formed was too weak to take the pressures involved in canning and prone to leaks. Double seams were only used for pails and cans of foods which did not need heating, for example, spices, baking powder, and lard. To guarantee the sturdy hermetic seal needed for food preservation, the double seam required some sort of sealant material, but this had proved difficult. People had already tried paper or thick rubber gaskets, but neither approach worked well.

The team that eventually solved the problem was centered on Max Ams, a German immigrant in New York who had started canning seafood for export in 1868 [4]. He believed his poor sales in Europe were due to his use of soldered cans, so he explored double seaming methods as a way to improve quality. His son, Charles, a chemist, developed an improved seal by dissolving rubber in ammonia to make a liquid that could be painted onto the curl, dusted with asbestos powder, and dried to form a thin layer. However, the sealant alone was ineffective unless the double seaming process could be scaled to industrial speeds.

In 1897, Ams hired his nephew Julius Brenzinger, a veteran of the German Army Engineering Corps, who quickly developed machines to apply the seal and dry them in rotary ovens. Brenzinger went on to invent a double seaming machine that worked by rotating the can on its vertical axis and pressing two chucks in succession to roll and press the end and body pieces together. No heat or solder was needed so the food did not burn, and because the entire can end could be sealed quickly after filling, there was no need for a filling hole.

Brenzinger's machines worked in his workshop, but moving them to the factory was not trivial. The workshop was staffed by the engineers who had built the new machines and who were dedicated to making them work. In contrast, the factory workers were only familiar with soldered cans, and they were much more interested in getting their food packed before it rotted than dealing with the intricacies of the temperamental new machines. What was needed was a dedicated cannery owner who had faith in the new technology, was willing to redesign their manufacturing process around it, and had the financial resources to persist until it succeeded.

William Bogle, a New York sales agent who knew Ams, had the industry connections and capital to support commercialization. In 1898, Bogle persuaded one of his customers, James Cobb—a canner in Fairport, New York—to experiment with the new Ams can. The new machines were awkward, and for several years the group struggled to adapt the prototypes so they would work in Cobb's cannery. They came close to giving up several times, but by 1903, the system was operational. In 1904, the five men involved, Max and Charles Ams, Julius Brenzinger, William Bogle, and James Cobb, went into business together as can-makers. As with the Nortons before them, improved transportation meant the business of making cans efficiently in a central highly automated factory could be physically distant from the businesses that filled them with food.

By the First World War, Ams and a competitor, the Bliss Company of Brooklyn, were making high-speed, fully automatic machines to manufacture and test metal cans. By this time, the familiar three-piece sanitary cans[3] had become the standard in most applications, although soldered cans continued to be used and improved well into the twentieth century. George Cobb, James' brother, looked back on his role in a technological innovation that changed the industry: "I have no predictions to make as to the final displacement of the Hole and Cap can by the Sanitary Can, each style of container has its particular advantages. I will, however, venture the opinion that the sanitary can will continue to show increased use commensurate with the expansion of the industry. It stands on the firm foundation of better canned foods for the consumer at lower costs to the canner" [5]. The same rationale of "better food" and "lower cost" motivates many of the technological innovations discussed in this chapter as well as the future hopes of modern canners.

4.3 Cooking Cans[4]

Leaving the process of filling the cans to the next chapters, the remaining problem common to all canned foods is cooking. The original technology here was the "three large copper kettles mounted in masonry on the furnaces" in Appert's workshop [6]. He would fill them with water and then add as many glass bottles as could fit before carefully heating the water to a boil. Cooking took a long time, sometimes up to 5 h, after which the water was drained, and another batch could be prepared.

Metal cans were initially processed in the same way as glass bottles, but, because they were more resistant to heat shock, they could simply be added to boiling water and removed when cooked. This apparently small change offered major energy savings and allowed the cooking to proceed as a continuous process. However, the cooking times were still very long. Any chef knows that cooking can be sped up by increasing the temperature, but the boiling point of water is fixed at 100 °C (212 °F). One might consider placing the cans directly over a flame, but the intense local heat would burn the food and soften the solder. The first practical solution to the problem of achieving a higher yet controlled cooking temperature was adding a salt, usually calcium chloride ("muriate of lime"), to raise the boiling point of water [7, 8]. There is some evidence this method was applied almost immediately by the group working on the Donkin patent in London in the 1810s, but, up to around 1840, the boiling

[3]The word "sanitary" was used to indicate the food was never in contact with solder. They were also sometimes known as open-top cans or simply as Ams cans.

[4]The development of can cooking methods in this period is described in Chapter 5 of Stuart Thorne's book [8], by Katherine Bitting in her 1937 history of the industry [19], and by the food technologists Samuel Goldblith [13, 14, 20] and Keith Farrer [7].

water was still the standard method [9]. The first patent for a salt bath was filed by Fastier in France in 1839, followed by a similar patent from Goldner and Wertheimer in England in 1841. By 1848, the Dangar brothers were canning meat in Australia using a brine bath, and Isaac Solomon introduced the method to American canneries in 1861.

Temperatures in salt baths ranged from 104 °C to as high as 127 °C (215–260 °F), reducing processing times from 4–5 h to less than an hour. Shorter, hotter cooking cycles generally produced higher-quality products, and the use of salts gave the canner more control over their process as they could now vary the type and concentration of salts (i.e., the temperature) as well as the cooking time. The salts did cause metal parts of the machines to corrode, but the main problem was the higher temperatures meant increased internal pressures and greater strain on the soldered seams.

The vent hole in the end, described in Chap. 3, served as a solution to this problem—but only if used correctly. Ideally, the tiny hole would be left open for the first part of the cooking and then sealed with a drop of solder after some of the trapped gases had escaped. In practice, however, this seemingly simple task proved difficult to perform quickly and consistently. Applying a precise drop of solder over a jet of steam escaping from a hot can was difficult, and if the can was allowed to cool to stop the steam, there was a risk of air being drawn back in.

Worse, if the can was inadequately vented, then the pressures achieved by further cooking at high temperatures could burst the soldered seams. *The London Illustrated News* described the consequences of such failures in 1852: "We well remember hearing recited to us the relation of an operator who was killed most ridiculously and ignobly by a boiled turkey. The canister in which the bird was soldered was exposed to the process of heating under pressure, and steam was generated beyond the power of the canister to endure. As a natural consequence of this the canister burst, and the dead turkey sprang from its coffin of tin plate, and, killing the cook, forthwith, made him a candidate for a leaden one, to illustrate in another way the effects of atmospheric exclusion" [10]. Some canners chose to cook at a lower temperature if they were not confident in the strength of their cans.

An alternative method to raise the temperature of cooking is to exploit the fact that water boils at higher temperatures at higher pressures. At the top of Mount Everest, for example, the atmospheric pressure is only about a third of that at sea level, and water boils at about 70 °C rather than 100 °C (158 vs 212 °F). Conversely, increasing pressure raises the boiling point. One straightforward, if somewhat risky, way to do that is to heat up a sealed container of water. Initially, the water boils at 100 °C, but as steam forms and pressure builds inside the container, more heat is required to maintain boiling. A modern domestic pressure cooker achieves pressures of about 1.6 atmospheres to get a cooking temperature of 115 °C (239 °F).

When a sealed can is cooked in a pressure cooker,[5] the pressure inside the can matches the pressure inside the pressure cooker so the seams are not stressed.

[5]The larger pressure cookers used for food preservation are called retorts (a name borrowed from the distillation apparatus used by early chemists) or sometimes autoclaves (from a French term for a self-locking cooking pot).

However, the pressure inside the pressure cooker is much higher than the pressure outside it, so its seams are stressed and can fail catastrophically if too much heating results in too great an internal pressure.

The first pressure cooker was the "bone digester" described by Denis Papin in 1681 and discussed in more detail in Chap. 7 [11]. It was a long iron pipe which formed a chamber, partly filled with the sample to be cooked along with some water to generate steam. The chamber was tightly sealed and heated with a few ounces of burning coals. Papin recognized the dangers of the process and, in some versions of his cooker included a weighted valve to regulate internal pressure. In other cases, users simply estimated the internal temperature by observing how long it took for a drop of water on the outside of the vessel to evaporate.

Nicolas Appert worked with large (400 l) Papin digesters in the 1830s, although he only used them to extract gelatin from bones and not to cook cans. In the 1831 edition of his book, he expressed the widespread public wariness of pressure cookers by telling the story of a Dr. Lemare whose carelessness with a version of Papin's digester had recently led to the death. Guiseppie Naldi, a popular opera singer: "Overconfident in his own skill, he recklessly trusted use of the [autoclave] to negligent and clumsy cooks but without giving them precise written instructions, which soon led to the most deplorable consequences. An awful accident caused the death of the unfortunate Naldi, chilling everyone with fright, and causing consternation in Paris and awakening all the old prejudices against the digester. People who had autoclaves hurried to get rid of them, and Mr. Lemare's useful invention was cursed by all" [12]. Appert even went so far as to patent a safety device, hoping it would help restore public confidence in using them.

After Nicolas Appert died in 1841, the unrelated Raymond Chevalier-Appert took over his canned food business and, in 1852, patented a retort designed for food preservation. This instrument had a built-in manometer for measuring pressure, a concave bottom to resist the high internal pressure, and a removable lid to allow pallets of food to be easily loaded with a hoist. However, it is not clear if it was ever successfully used. There were other patents, but the first practical retort for food preservation was developed by Albert Fryer in Manchester in 1870 [8]. Fryer's key innovation was an external steam generator which provided much more controllable pressure to the sealed chamber, rather than simply heating a sealed container of water. Fryer's device also allowed rapid cooling by introducing pressurized cold water to the retort at the end of the cook time. A similar retort fitted with a thermometer and a pressure gauge to regulate the process was developed independently by the canner Andrew Shriver in Baltimore in 1874 [13–15].

Retorts could reach temperatures similar to those of brine baths, with the added advantage that temperature could be precisely controlled by adjusting the pressure. Because sealed cans were heated inside a pressurized container, there was no stress on the soldered seams—allowing them to be fully closed immediately after filling, without the need for a vent hole. However, there were drawbacks. Setting up a brine bath required little more than a kettle and a sack of salt, while a retort was an expensive and potentially hazardous piece of equipment. As a result, retorts made more sense in operations handling large volumes of canned goods, favoring the growth of large, well-capitalized canneries.

4.4 A Hierarchy of Innovations

Nicolas Appert's preserved foods were a macroinvention (Chap. 2). He assembled existing technologies to do something radically different and publicized his result in a manner that other people copied and improve. The first great microinvention to follow was the metal cans which permitted more reliable processing and was better for naval expeditions (Chap. 3). The discoveries described in this chapter are all further microinventions.

If we attempted to arrange some of the microinventions discussed in this chapter into a hierarchy, we might instinctively put the Ams cans and the pressure retort at the top. Neither method was completely new—indeed double seams were a basic method in craft tinsmithing, and Papin's digester preceded practical use of retorts by almost 200 years—but a series of incremental improvements had brought both technologies to the point where they could be adopted for canning. While this chapter has focused on Papin's digester as a precursor to the modern canning retort, it was arguably more significant as a forerunner to the steam engine [16]. The extensive effort in the eighteenth and nineteenth centuries to build better steam engines, important in all parts of the new industrial economy, gave rise to many of the subtechnologies so useful for building a late nineteenth-century canning retort (e.g., steam valves, pressure gauges, efficient boilers, precisely manufactured parts). Similarly, the development of automated machines for folding and cutting tin for soldered cans advanced sheet metal processing to the point where it could be readily integrated into the Ams system once the innovation of the painted-on rubber seal was introduced.

There is a pattern to these major innovations that appears common in other industries. First, before the successful product was launched, various groups had already seen the potential of a new technology to solve problems with the existing way of doing things but failed to make it work (e.g., the paper and rubber gaskets used in Europe in double seams and Chevalier-Appert's retort). Second, when the breakthrough finally arrives, it frequently happens in several places at about the same time, often without any deliberate or clandestine sharing of ideas. In this case, Albert Fryer in Manchester and Andrew Shriver in Baltimore both developed useful retorts around the same time, and several individuals have been credited with inventing the salt bath. Joel Mokyr described this phenomenon in a discussion of the simultaneous invention of the blast furnace by two people in about the same period: "a 'problem' was defined jointly by a perceived market need and by the state of the art as defined by previous inventions and accumulation of knowledge. This 'problem' led ingenious men to explore various paths leading to a solution" [17]. The accumulated knowledge of pressurized steam technologies and the growing market needs of larger canning companies, a push and a pull, made the major innovation inevitable.

Further down the hierarchy of microinventions would be the various metal cutting, bending and seaming machines used to improve soldered cans in the nineteenth century. They were also improvements to an existing technology but are less important because each innovation led to a smaller change in industrial practice. However, many of these smaller changes were still significant enough to be described in patents, and the trade literature is full of advertisements for equipment claiming some advantage for a new product.

To take the argument even further, there is no reason not to extend the hierarchy of microinventions to include yet smaller changes to industrial practice. People working in industry will often share stories about clever fixes that someone rigged up to improve how things worked. For example, Condon Bush of the Bush Brothers Company in Tennessee was canning pork and beans.[6] The cans were cooked in a retort, but drops of water would condense on the lid and drip down the sides. If the operator failed to properly drain the retort, the cans at the bottom would sit in a pool of water and heat less quickly and less reliably. However, ensuring this simple task was being done consistently proved difficult. Bush solved the problem by placing a slotted metal rack at the bottom of the retort to lift the cans out of the water. Technologically, the solution was trivial—but in terms of solving a real problem, it was significant. These types of small innovations might be held as trade secrets, but surprisingly often the technical people at different companies will know one another socially and share stories. Bush told his friends in the business, and soon his modification became the accepted way to operate this type of retort.

Hidden away at the very bottom of our hierarchy are those innovations which were simply "bad ideas" at the time and place when they were tried and were quickly rejected. While this group is doubtless the largest in number—most ideas, after all, are bad ones—it will be hard to find documented or even remembered examples. Perhaps it includes those unfortunate individuals whose misuse of cooking technologies led to accidents or Chevalier-Appert's unused retort. It's important to remember this category of failed innovations exists, because success with technology depends as much on the prudence to reject bad ideas as it does on the courage to pursue good ones.

4.5 Reflections for Modern Technologists: We Are All Innovators

In the last chapter, I argued that the London canners—Donkin, Gamble, and Hall—are more deserving "heroes of the profession" for food technologists than the inventor of canning himself, Nicolas Appert. While they might lack the distinction of being the first, they made wise choices about how the technology could be improved

[6]This story was with me shared by Phil Perkins, a retired director at Bush Brothers.

to "make things work" in their commercial setting. Appert's practical skills and astonishing persistence may qualify him a hero for inventors, but process improvement in a practical context is the essence of what technologists do and that work deserves recognition and celebration. The examples in this chapter show how Donkin, Gamble, and Hall's great achievement with metal cans is part of a hierarchy of innovation which includes the work of technologists today.

The hierarchy includes people like Ams and the Nortons, who deliberately worked on transformative innovations, but also includes the unknown developers of the smaller machines that soldered seams or cut metal. It includes people much closer to day-to-day production—those who made systems work better through homemade solutions like Condon Bush's retort rack. It even involves the individuals who found better ways of doing things with existing machines: a better way to manage quality control data, a quicker way to move material around the factory, and a technique to clear jams on a fast-moving production line. From this perspective, no deliberate change in method or organization is too small to count as a technological innovation. The people who make these smaller innovations may not but famous, but they are often highly respected by their peers. They work mindfully on a process that matters to them, they see a better way of doing things, and their ideas are adopted and copied. Recognizing the role of innovator as an essential part of every career in technology is both empowering for technologists and, to some extent, democratizing for industrial progress.

Finally, as noted in the previous chapter, the evolutionary analogy often carries the almost moral assumption that new technology is intrinsically better. In reality, each new microinvention is a choice: adopt the new method or continue using the older one. In many cases, old and new technologies coexist within the same industry for years, with larger, better capitalized factories often leading innovation. Indeed, glass jars and water baths that would be familiar to Nicolas Appert are still used in home canning today.

Choosing to adopt a new process involves both costs—such as purchasing new equipment—and risks, including replacing a reliable method with one that might not work or missing out on other investment opportunities. Recall that the struggle to make Ams' cans work nearly costs James Cobb his business. On the other hand, choosing not to adopt a new technology carries its own risks, such as falling behind competitors who do realize its benefits.

In addition to creating innovations, technologists also play a vital role in evaluating which ones are worth pursuing. Ultimately, the wisdom of adopting a technology depends on local conditions. The next two chapters will explore these choices through examples drawn from the history of canning specific foods in specific places.

References

1. Unattributed. Tin cans and foods. *Sci. Am.* **48**, 341 (1883).
2. Unattributed. Norton's automatic can making machinery. *Am. Mach.* **26**, 1–2 (1883).
3. Stevenson, W. H. H. Cans and can-making machinery. In *A History of the Canning Industry by Its Most Prominent Men* (ed. Judge, A. J.). The Canning Trade, Baltimore, 92–93 (1914).

4. May, E. C. *The Canning Clan* (Macmillan Publishers Limited, New York, 1937).
5. Cobb, G. The development of the sanitary can. In *The History of the Canning Industry* (ed. Judge, A. I.). The Canning Trade, Baltimore, 94–96 (1914).
6. Appert, N., Bitting, K.G. (trans.): *The Book for All Households; or, The Art of Preserving Animal and Vegetable Substances for Many Years*. (1810)
7. Farrer, K. T. H. Who invented the Brine Bath? The Isaac Solomon myth. *Food Technol.* **33**, 75–77 (1979).
8. Thorne, S. *The History of Food Preservation* (Pantheon Publishing, 1986).
9. Cowell, N. D. *An Investigation of Early Methods of Food Preservation by Heat* (University of Reading, 1994).
10. Unattributed. The manufacture of preserved provisions. *Lond. Illus. News* **20**, 93–94 (1852).
11. Papin, D. *A New Digester or Engine for Softening Bones* (Printed by J. M. for H. Bonwicke, 1681).
12. Appert, N. *Le livre de tous les ménages* (Chez Barrois l'ainé, 1831).
13. Goldblith, S.A.: Thermal processing in retrospect and prospect. *Food Technol.* **30**, 32–33 (1976)
14. Goldblith, S. A. Controversy over the autoclave: who first used the autoclave for the preservation of foods?. *Food Technol.* **26**, 62–65 (1972).
15. Petrick, G. M. An ambivalent diet: the industrialization of canning. *OAH Mag. Hist.* **24**, 35–38 (2010).
16. Needham, J. The pre-natal history of the steam-engine. *Trans. Newcom. Soc.* **35**, 3–58 (1962).
17. Mokyr, J. The later nineteenth century. In *The Lever of Riches: Technological Creativity and Economic Progress* 370–386 (Oxford University Press, 1992). https://doi.org/10.4324/9781315562315-22.
18. Pearson, G. S. *The Democratization of Food: Tin Cans and the Growth of the American Food Processing Industry, 1810–1940* (Lehigh University, 2016).
19. Bitting, K. G. A benefactor of humanity. In *Reproduced in: Appertizing or the Art of Canning; Its History and Development* (ed. Bitting, A. W.). The Trade Pressroom, San Francisco, 7–18 (1937).
20. Goldblith, S. A. A condensed history of the science and technology of thermal processing – part 1. *Food Technol.* **25**, 1256–1262 (1971).

Chapter 5
Canning Salmon from the Pacific

Abstract The Pacific Northwest 1862-1905. Managing labor and technology for a canning season that lasts only a few weeks. The replacement of labor by technology is not an inevitable law but depends on local circumstances and individual choices. Race, labor, and technology.

5.1 Canning a Particular Food in a Particular Place

The previous chapter showed how two portions of the canning operation, making cans and cooking cans, were industrialized through a series of microinventions. Over about a century, can manufacturing evolved from a system of craft manufacturing to integrated systems of machines arranged in a production line. Over the same period, salt baths replaced water baths for cooking until they were in turn replaced by pressure retorts. The speed of production increased, while the costs of products decreased. But this is only a part of the story—people, food, and place all matter too.

First, despite the replacement of many jobs by machines, the need for labor is never fully eliminated. Even the most modern canning machines need at least some human attention—to clear jams, to adjust settings, and to switch over between protocols. Human workers are more flexible than machines, so for certain parts of the canning process—especially those involving irregular foods or infrequent operations—they remain essential.[1] A cannery can only operate in a location where it can attract sufficient workers who are willing accept the wages offered. Furthermore, people have preferences in the work they do, and, when they have the power, they use it to make the work more to their liking—either by leaving for alternative employment or by insisting on improved pay and conditions.

Second, most fresh foods have a very short shelf life—this was the entire motivation of preservation in the first place—so the cannery must be located near to where the

[1] I recall visiting a highly automated cannery which occasionally made pork and beans. When they did, it was one worker's responsibility to drop a small piece of fatback bacon into each can by hand. This low-tech arrangement made more sense than building a machine for a single, occasional task.

J. Coupland, *Containing Nature*, https://doi.org/10.1007/978-3-032-11882-0_5

plants are picked or the animals slaughtered. How near that should be depends on the nature of the food—how quickly it spoils—as well as how effective the transport system is. In addition, most of the foods which are canned have a harvest season when machines and sufficient labor must be available to process everything before it spoils. Unless the canner can find other things to pack in the off-season, the equipment will sit idle, while the interest on the money borrowed to purchase it continues to accumulate.

Because of these two factors, there is a trade-off between building large, automated canneries in a central location near population centers and transportation hubs and smaller, less efficient canneries closer to the farms. Similarly, there is trade-off between the efficiency gains from specializing in a single product and the flexibility of being able to can many different foods.

Finally, there is history. People live and work in places and become attached to them. A farmer starts a small cannery to pack some of their own harvests. Every few years, they buy new equipment or extend their buildings to meet the most pressing needs of the growing business. A generation later, the factory is an accumulation of technologies and practices with its own idiosyncrasies and inefficiencies they have learned to manage. The next generation of owners can imagine a more efficient cannery—perhaps a new building close to the highway with the latest equipment—but their history means they are probably going to continue to modify what they have inherited. It would simply cost too much, socially as well as economically, to move. Technology implementation rarely occurs on a blank canvas—choices made in the past can determine the best choice today.

Some of these factors apply to all industrial operations, while others are particular to canning. Taken together, they act as constraints on the simple "bigger, faster, more efficient" logic of industrialization. Effective technology work involves balancing these factors and making the process work in a particular local context. In the next two chapters, we will see two historical examples of how this was achieved in the canning industry—salmon on the Pacific Coast and soup by the Campbell's company in New Jersey. While any industry segment would have provided its own insights, these two stories provide examples of small vs. large canneries, a poorly capitalized vs. a well-capitalized industry, a region with an inadequate vs. ample labor supply, and seasonal vs. year-round operation.

5.2 Canning Salmon on the Pacific Coast[2]

In October 1805, while Nicolas Appert was working on canning in Paris, an expedition of Americans led by Meriwether Lewis and William Clark reached the mouth of the Columbia River on the border of present-day Washington and Oregon. The

[2]The key histories of canning technology in this region and period are by Duncan Stacey [17], Keith Ralston [19], and Patrick O'Bannon [12, 13]. Dianne Newell [20], Alicja Muszynski [21], and Chris Friday [14] focus on labor organization in the canneries. These sources are the basis on much of this chapter.

river teemed with fish. According to Rev. Gustavus Hines, a Methodist missionary who worked in the region in the 1840s: "They literally fill the rivers of Oregon, in their season. And at all the falls and cascades in the various rivers of the country, the quantities taken and that might be taken, are beyond all calculation. As they penetrate far into the interior, they afford almost inexhaustible supplies to the Indian tribes of the country, as well as the whites, many of whom depend almost entirely upon such supplies, for the first year, after settling in the country" [1].

Just as the plains' tribes depended on the buffalo, the Indians of the Pacific Northwest had historically depended on salmon as a food source for at least the part of each year. They caught fish with hooks, nets, and traps and either ate it fresh or preserved it by smoking or by drying and pounding.[3] Some scholars have argued that the tragic reason the fish were so abundant in the latter part of the century was precisely because Indian fishing had collapsed due to the arrival of human epidemics brought by Europeans [2].

The migrants who followed Lewis and Clark in the nineteenth century also ate the fish, but they also were part of the global trading economy and were in the region seeking to extract value from the environment—things they could transport and sell. In the earliest days of European settlement, the Hudson Bay Company traded with Indian people for salmon, which they salted and packed into barrels for trade around the Pacific basin and particularly to Hawaii. However, anything of value from the Pacific Northwest intended for Europe or the big cities of the East coast had to be shipped around Cape Horn. Fish spoiled on the way, so the only marine products that could be exported were pelts and oils. Canning changed everything.

Small amounts of Atlantic salmon were canned in the eastern United States and Canada in the 1840s, and some of the canners started to migrate west in the 1860s, attracted by reports of the abundance of the salmon runs there. The first salmon cannery in the Pacific West was built in 1864 by the Hume brothers, fishermen from Maine, along with their friend Andrew Hapgood, a tinsmith who had some experience canning Atlantic lobsters. The cannery was on a houseboat and a shack on the riverbank near the downtown area of modern Sacramento.[4]

The methods they used were very basic and probably comparable with those Donkin, Gamble, and Hall had used in London in the 1810s. In his romanticized

[3]Washington Irving described the Indian processes of salmon preservation from firsthand reports he had heard. "They [the fish] are then cured and packed in a peculiar manner. After having been disemboweled, they are exposed to the sun on scaffolds erected on the riverbanks. When sufficiently dry, they are pounded fine between two stones, pressed into the smallest compass, and packed in baskets or bales of grass matting, about two feet long and one in diameter, lined with the cured skin of a salmon. The top is likewise covered with fish skins, secured by cords passing through holes in the edge of the basket. Packages are then made, each containing twelve of these bales, seven at bottom, five at top, pressed close to each other, with the corded side upward, wrapped in mats and corded. These are placed in dry situations, and again covered with matting. Each of these packages contains from ninety to a hundred pounds of dried fish, which in this state will keep sound for several years" [22].

[4]The story of the Sacramento cannery is told best by Collins [23].

history of the industry, Earl Chapin May described their initial equipment as "a screw hand press, a set of cast iron dies for making can tops, another set for making can bottoms, a set of squaring shears to cut out the 'bodies', and one set of rotary shears for out the round ends, bench shears, and snips, …, [a] pair of twenty-four inch rolls for making the bodies cylindrical, …, an anvil weighing fifty pounds, a forging hammer, and a tinners hammer, …, a set of punches, …, a rivet set and a grooving set, two iron slabs with grooves in which to mold strips of solder, an iron clamp to hold the bodies while the seams were being soldered and a triangular piece of iron to assist in holding the bodies" [3].

The salmon were cleaned on the moored boat and then brought into the shack and cut into pieces by hand, using a stick to judge the appropriate length of fish for the can. The salmon steak was pushed into the can by hand along with a little salt, boiled in fresh water for an hour to exhaust the air so they would form an internal vacuum when cooled, and then closed sealed with solder. They were then fully cooked in a brine bath at 108–110 °C (226–230 °F), painted with a red lacquer to prevent rust, labeled, and packed [4]. The work was organized as a production line with tasks separated between workers, but each task was craft labor using only a few hand tools.

The initial quality of the canned salmon was very poor, with over half the cans failing due to poor seals, and the product they did make was not popular locally where the population was sparse and fresh fish abundant. Still, they packed 2000 cases (the standard size for a case was 48 lbs of fish, probably packed in 1 lb cans) in their first year and found export markets in Australia, New Zealand, and Latin America and, by the 1870s, England. They had demonstrated Pacific salmon could be canned and traded globally for a profit. By the end of the century, canned salmon was selling in England for less than half the price of fresh meat and had become an important part of the diet of the poor.

The Hume-Hapgood cannery on the Sacramento River shut down after drought and mine runoff gave the fish a poor taste—and because the boat they had used for processing sank. The group moved north and, in 1866, set up on the Columbia River where the salmon runs were larger. Other canneries soon followed [5]. Then, as the fish declined there, the canners continued moving their operations northward, abandoning one river for another. It may not be fair to blame the decline of salmon solely on the canners alone; some estimates suggest indigenous people took about as many fish before European contact [6]. More likely, the other things the new migrants did to extract value from the landscape—runoff from mining, dams, irrigation projects, dredging—destroyed its capacity to support such large numbers of fish. The native peoples who had depended on the salmon for long before the canners arrived did not move, and their descendants fish the same waters today [7].

Salmon canneries spread north from the Columbia into Canada in the 1870s and, by 1878, Alaska. By the start of the twentieth century, Alaska had grown to become the center of the salmon canning industry, while the more southerly fisheries, and especially California, were relatively minor backwaters. The minimal capital needed for the first wave of new enterprises tended to be from brokers in San Francisco, Victoria, or Vancouver who would underwrite the costs of the cannery for the

18 months between sending an order for English tinplate to selling the first packed salmon [8]. Later in the century, banks and limited corporations organized as far away as London provided the money for larger and more consolidated cannery companies, but the first canneries were shoestring operations. They were also quite small, not only because of their poor capitalization but also because they only needed the capacity to process the fish brought ashore in their immediate area.

The canneries opened a few weeks before the salmon were expected, with a small crew to make cans and to make sure the equipment was ready while the rest of the workers, perhaps 130, were brought in closer to the time when the salmon were expected. The run—the only time the cannery could make money—might only last for a few weeks or months and halted completely by late summer.

Because refrigeration and transportation were very limited, and because the fish spoiled so quickly, canneries tended to be located close to the salmon rivers and were therefore often very isolated. In many cases, the only way to access the remote canneries was by sea, so all the supplies, personnel, fuel, and equipment had to be brought in by boat, and bunkhouses and kitchens were provided for the workers. Assembling the resources needed, especially workers, in remote locations at specific times was a problem characteristic of Pacific salmon canning.

5.3 Workers for the Canneries

The pioneering Hume-Hapgood cannery in Sacramento was a small family business with no real distinction between management and labor. The Hume brothers fished and worked in the cannery alongside a few men they hired from local farms. When they moved north to the Columbia, their greater needs put strain on the local labor supply so that, by 1868, they were forced to advertise in Portland newspapers for "men and boys" to work in the canneries. These new workers came from a range of ethnic backgrounds often different from those of the company founders and created a clear division between management and labor in the factories. In California, one of the sharpest divisions was racial.

Chinese people had begun to immigrate to the United States early in the nineteenth century, but their numbers increased dramatically around 1850 as men from Guangdong arrived looking for work in the mines, in agriculture, and in railroad construction [9, 10]. They experienced intense racial discrimination and even lethal violence from local whites who nonetheless eagerly exploited them as a cheap, mobile, and capable source of labor. As railroad work declined in the 1870s, many Chinese laborers were recruited by the growing canneries.

About 10% of Hume-Hapgood's 1870 crew on the Columbia, 13 workers, were Chinese—the first recruited by a California cannery. To find the new workers, George Hume had depended on Sam Mott, a Chinese cook, who served as a labor contractor. Mott had language skills and personal contacts that allowed him to act as a liaison between the English-speaking canners and the small local Chinese community. The combination of European-American owners and Chinese workers and labor contractors would define Pacific salmon canning for the rest of the century.[5]

The Anglo-Indian writer Rudyard Kipling made a brief visit to a cannery near Portland, Oregon, in the late 1880s and in a few paragraphs captured both the nature of the process and the racial attitudes at the time: "The steamer halted at a rude wooden warehouse built on piles in a lonely reach of the river, and sent in the fish. I followed them up a scale-strewn, fishy incline that led to the cannery. The crazy building was quivering with the machinery on its floors, and a glittering bank of tin-scraps twenty feet high showed where the waste was thrown after the cans had been punched. Only Chinamen were employed on the work, and they looked like blood-besmeared yellow devils, as they crossed the rifts of sunlight that lay upon the floor. When our consignment arrived, the rough wooden boxes broke of themselves as they were dumped down under a jet of water, and the salmon burst out in a stream of quicksilver. A Chinaman jerked up a twenty-pounder, beheaded and de-tailed it with two swift strokes of a knife, flicked out its internal arrangements with a third, and cast it into a bloody-dyed tank. The headless fish leaped from under his hands as though they were facing a rapid. Other Chinamen pulled them from the vat and thrust them under a thing like a chaffcutter, which, descending, hewed them into unseemly red gobbets fit for the can. More Chinamen with yellow, crooked fingers, jammed the stuff into the cans, which slid down some marvelous machine forthwith, soldering their own tops as they passed. Each can was hastily tested for flaws, and then sunk, with a hundred companions, into a vat of boiling water, there to be half cooked for a few minutes. The cans bulged slightly after the operation, and were therefore slidden along by the trolleyful to men with needles and soldering irons, who vented them, and soldered the aperture. Except for the label, the "finest Columbia salmon" was ready for the market. I was impressed, not so much with the speed of the manufacture, as the character of the factory. Inside, on a floor ninety by forty, the most civilised and murderous of machinery. Outside, three footsteps, the thick-growing pines and the immense solitude of the hills. Our steamer only stayed twenty minutes at that place, but I counted two hundred and forty finished cans, made from the catch of the previous night, ere I left the slippery, blood-stained, scale-spangled, oily floors, and the offal-smeared Chinamen"[6] [11].

While outsiders like Kipling tended to see the Chinese workers as undifferentiated and alien, there were important hierarchies within the work crew based on job

[5] Some plants, particularly those in Alaska and Canada, depended heavily on the labor of native women and children who would often make their camps near the cannery. As the number of Chinese workers available declined toward the end of the century, Chinese labor contractors increasingly organized Mexican, Filipino, and Japanese work crews, who were fed and housed separately in the canneries.

[6] By 1889, the boiling water would likely have had salts added to raise the cooking temperature, or there may have been a retorting process that Kipling missed during his short visit. The "thing like a chaffcutter" was probably a gang knife—a series of linked, evenly spaced blades that could be pulled down to make simultaneous cuts along the length of the gutted salmon and produce steaks sized for the cans.

roles. The "China boss" served as a foreman for the Chinese workers and enjoyed a great deal of power in the cannery. He could also be a target for worker anger if wages went unpaid or if the living conditions were inadequate.

Below him, the highest paid and highest status workers in the factory were the butchers (responsible for removing the heads, tails, and fins and cleaning the cavity). The salmon butchers were the elite of the cannery workers. Fish would be dumped, thigh-deep, onto the cannery floor where a team of butchers would haul them up onto tables and process them with eight precise cuts. A good butcher could process three fish a minute and maintain that pace for ten or more hours per day for weeks on end until the rush of salmon slowed [12, 13].

Butchers were always in short supply, as not everyone wanted the job. It was among the most physically demanding and dangerous work in the cannery. They also had to start early to build a buffer of processed fish, allowing the rest of the production line to keep moving. Despite this, some butchers tried to restrict access into their profession by limiting opportunities for informal apprenticeships and by keeping the tricks of the trade to themselves. They would modify their tools to better meet their needs, filing down the long knives to shorter, more useful blades ("because you can't afford to be swinging a big knife around") and sharpening them carefully to slice through flesh without dulling on the bone [14].

If we think of technology as the ways people work with the material world to better achieve their goals, then the butchers carried a form of technology in their muscles, knives, and brains—a technology the cannery owners had to pay a premium to use. Chinese salmon butchers were paid between 20% and 50% more than other workers, and even more when they were scarce.

Lower in status and pay than the butchers were the cappers (who were responsible for quickly soldering the ends onto the cans after they had been filled) and testers (who tapped the cooked cans with a nail and listened for the dull sound indicative of a leak; there were very few testers needed, and their light work was highly sought after and often went to men with good connections to the labor contractor). Mid-tier work roles included the slimers (who cleaned out the fish cavities with cold water, brushes, and knives; it was the wettest job in the plant) and the fish pitchers (who used "pews"—long pitchforks—to load the heavy salmon into the cannery by spearing them through the heads; extremely dangerous work as they had to wade among masses of slimy fish). The lowest tier jobs included loading and unloading the retorts, washing the cans in caustic lye to remove grease, painting them with lacquer to prevent rust, and labeling.

The racial and gender assumptions of the time meant some tasks in their nature required a white man whose work deserved higher pay [15]. High-status white workers in the plant might include an engineer responsible for the operation of the retort and the bookkeeper, while poorer white people were sometimes also hired to perform lower status tasks or to take the pressure off the Chinese crew during a rush. These workers did not report to the China boss.

In the early days of the industry, the Chinese were just one group of workers among many, but in time they came to be seen as the "natural" choice. Part of this reason was simply convenience. By negotiating with a labor contractor, a canner could arrange all the workers he needed for a season just as easily as he would order tinplate from another supplier and shipping contracts from a third. The labor contractors would recruit the workers, handle payroll, and provide a foreman (the "China boss") to supervise the work crew in the cannery.

Beyond simple convenience though, racial ideologies of the time made it easier to believe that different peoples possessed inherent traits that suited them for specific roles in life. While white cannery owners found it easy to believe that white people were natural leaders, they also believed that Chinese men were naturally suited to the work of canneries. The owners would approvingly describe their workers as naturally "docile" and "hardworking," although their behavior actually reflected the fact that they were often in debt to the labor contractors and had limited support available beyond their isolated work crew. They had little choice but to accept whatever deals they could get and then work hard. Sociologist Alicja Muszynski also argued that cannery owners tended to feminize Chinese men because of their smaller bodies and different appearance and so saw them as suitable for traditionally female work roles such as laundry and cooking—hence the indoor work of the canneries rather than the outdoor work of the fisheries [16].

If we take the "baseline" of the canning process as the one Kipling described, then any variation in methods had to be expected to make more money than that. Broadly speaking, any change that reduced the cost of production was an improvement, but the trade-offs described above meant the adoption of technology was neither linear nor inevitable. With this in mind, we can examine how technologies were adopted in the Pacific Northwestern salmon canneries.

In general, technologies involved in the processing of tin cans were developed in the East by other segments of the canning industry and then modified to the local demands of canning salmon. In contrast, technologies involved in processing fish were developed locally and were intrinsically more difficult because fish are themselves so variable.

The first wave of technological developments in Pacific Northwestern salmon canneries (up to about 1900) were about trying to make the workers more efficient. The second wave, in the twentieth century, sought to fully automate the factories and eliminate their dependence on skilled craft labor, particularly the salmon butchers. Both waves of mechanization provoked worker resistance when livelihoods were threatened.

5.4 Late Nineteenth-Century Innovations in Salmon Canning Technology

Some technologies found a ready home in the salmon canneries. The pressure retort, pioneered in Baltimore in about 1874, spread very quickly to the Pacific Coast. By 1880, and despite Kipling's report, it was very unusual to see the open pans being used to process cans. The retorts were changed to handle the rush of the salmon run.

They were rotated so that their doors were on the side instead of the top. This allowed workers to load trays of cans onto carts and easily roll them in and out.

There was little widespread worker resistance to the introduction of retorts, perhaps because the increased speed of production and improvements in quality made everyone more money. Additionally, cannery workers responsible for cooking the cans were among the lowest paid and lowest status in the factory, and they could often find work elsewhere without much difficulty.

Other canning technologies developed in the East were harder to apply in the Pacific Northwest. As we saw in Chap. 4, can making had shifted from hand manufacturing—often in the canneries during the off-season—to large, highly automated factories located on rail lines that could supply many canneries with preformed cans at a low cost. The first of these factories was opened by the Nortons in Chicago in 1883, and, soon afterward, the brothers contracted with a company to manufacture cans for the salmon canneries in Alaska.

However, what had worked so well in the East did not transfer easily to the Pacific Northwest. Most salmon canneries were accessible only by sea, and transportation was expensive. Sheets of tinplate took up less space than finished cans, making them cheaper to ship. Additionally, seasonal Chinese work crews preferred longer contracts, and canneries that offered a few relatively quiet weeks of can making before the salmon arrived were more attractive. When labor was hard to find, this could be a decisive factor. Consequently, Pacific salmon canners did not immediately adopt the new premade can technology, rationalizing their rejection of "progress" by arguing making cans was not that expensive anyway and the machine-made cans might even be inferior.

Eventually, assessments of can-making machines shifted as the technology improved and labor conditions evolved. In some cases, the machines were installed directly in salmon canneries to reduce transportation costs and to ease labor resistance by preserving local can-making jobs. As always, local conditions played a key role in how technology was adopted.

The first widespread use of premade cans occurred in Alaska in the mid-1880s, where canneries were larger and better capitalized and faced particularly severe logistical challenges in maintaining large work crews in remote locations. Even getting a crew to an Alaskan cannery early enough to make cans before the salmon run could be impossible due to bad weather. The situation was different in British Columbia and the contiguous United States, where varying labor and economic conditions led many canneries to continue using the older, handmade can technology well into the twentieth century.

The preferences of the work crews was also a factor slowing the adoption of machines to solder the ends onto cans after filling. Eastern canners had implemented a range of sophisticated soldering machines, including the Howe floater (Chap. 4). The logic of industrial efficiency was compelling: a skilled worker could close about one can a minute by hand, while a Howe floater operated by two less skilled workers could close 45–50. While a 20-person soldering team might cost the owner of a salmon cannery $1000 per month, the two operators of the machine might be hired for $50 for the same period. The machines themselves were relatively cheap (~$250–$300), and they used less solder, so they would seem an obvious technology to adopt quickly.

However, unlike the retort, introducing a Howe floater would displace skilled workers. Owners stood to benefit from higher productivity and lower payroll costs, but some of the higher-paid, higher-status workers would lose their jobs. In Eastern canneries, cappers organized into unions and tried to resist, but they were quickly replaced by machines. In the salmon canneries, cappers were less organized, but they held significant power: any labor disruption during the few critical weeks of the salmon run could ruin the cannery's profitability for the entire year. The Howe floater was first tested in an Oregon cannery in 1877, almost as soon as it had been invented, but workers successfully resisted the adoption of capping machines until the mid-1880s.

Beyond these imported innovations, there were some locally designed machines developed to handle fish processing more quickly. A good example is the process used to slice the cleaned fish into appropriate lengths to fit into the can. The original method used on the Sacramento River was a sharp knife and a measuring stick, but by the 1880s, this was replaced by a hand-operated "gang knife" like the one Kipling saw in Oregon. A pair of gutted fish were placed into a cradle, and a worker would struggle to pull down a hinged knife with multiple blades spaced to cut the fish into three or four pieces, each the length of the can. This was exceptionally difficult work, especially when packing for half pound cans when the knife had more blades to cut the fish into smaller pieces and the machine frequently jammed.

By 1899 the task was made easier by using gang knives with rotating blades that could be operated by turning a handle to slice the fish, and later still steam power was supplied to the cutter [17]. These machines served to "standardize" the variable shape of the fish into the fixed dimensions of the can but first required the head, fins, and guts to be removed by the skilled hand work of the salmon butchers, whose skills remained essential to cannery operations.

Other machines developed locally to shape salmon for cans were less successful. In 1883, Mathias Jensen, a fisherman on the Columbia River, invented a machine to slice meat from the fish and force it into cans. Once again, the logic of industrial efficiency was compelling—even the first models could fill an impressive 50 cans per minute—and here the labor displaced was unskilled and so could be reassigned to other cannery tasks. In this case, the resistance to adoption of the technology came from outside the cannery.

English consumers had grown used to a can of salmon being essentially an intact salmon steak pressed into the can, with visible bones and intact muscle tissues. The Jensen filler, however, produced cans filled with shredded pieces of fish whose "mangled" appearance was seen as inferior. Canadian canners, who relied heavily on export sales to England since tariffs made their product uncompetitive in the United States, quickly reverted to hand-filling salmon steaks to meet consumer expectations. In contrast, the highly mechanized Alaskan canneries readily adopted the machine filler as it allowed them to compete more effectively on price in the American market, where people were willing to accept the shredded pieces.

Hand-filled cans continued to be used well into the twentieth century for premium products, especially if cheaper labor was available. Once again, the "better" technology wasn't better for everyone—whether it was adopted or not depended on local circumstances.

In the early days of Pacific canning, a cannery might pack about 240 cases of fish per day. By 1883, a typical cannery with a crew of 130–150 used soldering machines and retorts to pack 1000 cases. By the end of the century, larger crews but with fewer skilled workers were packing 2400 cases a day [12]. The last remaining craft operation was the work of the famous Chinese salmon butchers, and how this was eventually automated brings the conflicts between technology and craft labor in salmon canning into the sharpest focus.

5.5 Early Twentieth-Century Innovations in Salmon Canning Technology

Chinese immigration into the United States was outlawed in 1882, and, while it remained legal in Canada until 1923, immigrants there had to pay a special fee. By the start of the twentieth century, the Chinese workers who had dominated Pacific salmon canning since the 1870s were aging or dying or had returned to China. Mechanization had reduced the need for labor to some extent and eliminated some of the skilled jobs while increasing the overall speed of the process—but the task of manual fish butchery remained.

The only way canners could increase factory throughput was by hiring more butchers, yet they were increasingly rare, which gave them even greater leverage within the cannery. Even when butchers could be found, they were expensive and often found to be "troublesome" for the managers. In his history of canning technology on the Fraser River in Canada, Duncan Stacey records cannery laborers who had earned $35–50 a month in 1901 making not less than $65 in 1908. Even then they could not "pack to capacity" because of a lack of Chinese workers [17]. The butchering operation had become a bottleneck.

Several individuals recognized the opportunity and began experimenting with fish butchering machines around the turn of the century. The key breakthrough came from a Canadian working in Seattle—Edmund Smith [13]. Despite having no engineering background, Smith began working on his machines in 1902 and had a prototype working by 1903. After observing its performance during that season, he redesigned the machine with a smaller footprint to fit more easily into canneries and had a commercial version ready for the 1904 salmon run.

As Max Ams was learning with his double seaming machines in New York at the same time (Chap. 4), equipment that performs well in a workshop often needs considerable adjustment function reliably under real production conditions. Rather than trying to sell his prototype machines outright, Smith leased them to canneries for a few cents per case packed. He sent engineers out with each machine to supervise installation and to keep it running effectively. If the machine ran well, then both Smith and the canners who leased his machine made more money. More importantly, having his engineers out in the field allowed Smith to treat his first season as a form of "beta testing." He learned from his engineers' experiences, made improvements, and had his refined machines ready for sale in time for the 1905 salmon run.

The operation of the mechanical fish butcher was described in the 1911 novel "The Silver Horde": "It is an awkward-looking, yet very effective contrivance of revolving knives and conveyors which seizes the fish whole and delivers it cleaned, clipped, cut, and ready to be washed. With superhuman dexterity it does the work of twenty lightning-like butchers" [18]. Workers first used a band saw to cut the heads and tails from each fish, a job which paid less than half what a butcher would make. Next, they lifted the fish onto a vertical wheel which brought it through a series of fixed knives and brushes that sliced off the fins, opened the body cavity, and scraped out the viscera. The fish carcass then had to be washed by hand to remove any remaining blood and guts, before it could be cut into steaks with gang knives and pressed into cans.

The original machines cost $2600, required three operators, and processed 40 fish in a minute. It was not a complete automation of salmon butchery, but it didn't need to be. Its success lies in enabling canners to replace the expensive and increasingly scarce skilled salmon butchers with less skilled workers who were cheaper to hire and easier to train. For decades, the butchers had long held power in the canneries because of the "technology" they carried in their hands, their knives, and their brains—technology they controlled—and cannery owners had to pay to use. The Smith machine meant the control of the fish butchering technology had passed to the cannery owners.

Advertising emphasized the machine as a direct replacement for labor, specifically targeting the displacement of Chinese butchers. Patrick O'Bannon cites a testimonial from a cannery superintendent reprinted as an advertisement: "I had less trouble with the Smith Cleaning Machine than with any other piece of machinery I had in my cannery…Hereafter I want no Chinese butchers in the plant if I can help it" [12]. The relationship between the technology and the power structures in the canneries is reflected in the name given to the machine—the Iron Chink.

The Chinese workers themselves opposed the introduction of the mechanical butcher and, in one case, threatened the life of a white worker charged with installing it.[7] However, the Chinese labor gang bosses did not necessarily object, as they could more easily assemble crews without butchers. The gang bosses were typically paid a little less for the crews without butchers, but the lower wage costs and higher

[7] The status of Chinese workers as "others" in the canneries at this time is captured in this September 1905 report on an instance of resistance to the introduction of a mechanical butcher in Washington State: "When Fred Ohmert, a mechanic in the employ of the cannery adjusted the machine and made it work with great rapidity, the Celestial laborers were very wroth and threatened Ohmert with extinction. A careful watch was kept of them after that and they did not carry out their threats" [24]. The unnamed reporter, who may have been Ohmert himself, did not think to ask the workers their views; they were simply watched with suspicion as if they were prisoners.

overall productivity increased the labor contractor's profits just as they did the profits of the cannery.

In the early days of the industry, most canneries were small, as the range over which fish could be transported was limited. However, in the early twentieth century, reliable gasoline engines for boats and improvements in refrigeration made it easier to concentrate fish from a wider area into larger, consolidated canneries. Where the volume of fish was greater, particularly in Alaska, it made more sense to invest in expensive equipment to process them, and butchering machines were quickly adopted. They offered much greater speed and could even support multiple production lines in the same factory, allowing different species of fish and different styles of can to be processed at the same time.

In other places, local circumstances meant the mechanical butcher did not make sense. For example, canneries specializing in the larger chinook salmon had to wait until 1917 until a machine big enough to accommodate them was developed. In places where the volume of fish was smaller, the butchering machines were as useful. For example, there were no butchering machines on the Columbia River in 1937. Once again, the better technology wasn't better for everyone, and where it was adopted depended on local circumstances.

5.6 Reflections for Modern Technologists: Making Choices

The simplest narrative of industrialization is the replacement of skilled craft processes by automated systems of machines. In Chap. 4, we explored how this process of industrial development can be looked at as evolutionary, with small modifications ("microinventions") serving as mutations and "efficiency" serving as the selection pressure. The transformation of Pacific salmon canning fits this pattern in broad terms.

However, the analogy can take on the appearance of an inevitable natural law. In reality, the changes that now seem inevitable were, at the time, real decisions made by real people. The constraints imposed by the people, the food, the environment, and the history of what went before meant the "better technology" wasn't better for everyone. Choices had to be made.

The choice to adopt a new technology can never be completely rational—it always involves a judgment about future conditions. Edmund Smith claims miraculous performance for his new salmon butchering machine—but can you trust him? Even if the machine works as promised, will it integrate smoothly into your existing cannery operations? Can you find workers to operate it, or spare parts if it breaks down? It might seem wise to wait a few years and learn from the early adopters. On the other hand, your labor contractor couldn't find enough Chinese butchers last year to keep up with the run. Choices have to be made.

The choices about technology are mirrored in the choices that must be made about labor. The simple narrative of industrialization treats labor as a replaceable part which will inevitably be deskilled and eventually replaced as the machines become more sophisticated. In reality, what we call "labor" are people with their own preferences and who may have power to influence the way a factory works. If the labor supply is limited—either by the action of a union or through simple scarcity of people—workers' power increases.

Finding a suitable crew was a particular challenge for cannery owners in the Pacific Northwest. If offering a few extra weeks making cans could help you secure the workers you need, then perhaps those new premade cans aren't worth it this year? If you had a good relationship with your labor contractor and the crew had worked well the previous year, you might decide to stick with the status quo, but if you had had to throw fish away last summer because the butcher they recruited was unreliable, or if you believed the river was due an exceptionally large salmon run next year, then you might spend the winter browsing equipment catalogs. The right decision depends on your expectations of the future.

Future conditions were especially hard to predict in the salmon canneries. If the plan was to simply can corn or tomatoes, it would be relatively easy to walk around the surrounding farms and talk to farmers. How much crop had been planted? How well was it growing? It might even be possible to sign a contract with farmers in advance to grow the varieties you want and sell them to you on a given date at a given price. Knowing all this, the vegetable canner could prepare for the harvest volume by carefully selecting a combination of labor and machines.

Salmon are much less predictable. A folk wisdom at the time suggested sockeye salmon would have one excellent year followed by a good year followed by two poor years, but, in fact, no one really knew [19]. Even within a season, the number of fish available to be packed on a given day would increase, reach a peak, and then decrease again. With little capacity to keep fish for much longer than a day, did the cannery need capacity to pack for those few peak days or would some level of waste be tolerable?

Choices about technology and labor had to be made well in advance, and the remoteness of the canneries meant there was little opportunity to make changes mid-season. As the historian Patrick O'Bannon put it, "Like a gambler, or a general, the salmon canner studied the signs and laid his plans. Then he waited and trusted to luck" [12]. Modern business leaders may use sophisticated economic models to hedge against risk, but the right decision ultimately depends on judgment rather than any reliable rule. Good judgment stems from experience—and experience can be cultivated deliberately. While being involved with the business over time and encountering a variety of conditions are certainly important, expanding your "experience network" can be just as valuable. By connecting with a local network of peers engaged in similar work, you can learn from their experiences as well as your own.

References

1. Hines, G.: Wild Life in Oregon. Worthington Co. (1887)
2. Schalk, R.F.: Estimating salmon and steelhead usage in the Columbia Basin before 1850: the anthropological perspective. Northwest Environ. J. **2**, 1–29 (1986)
3. May, E.C.: The Canning Clan. Macmillan Publishers Limited (1937)

4. Hume, R.D.: The first salmon cannery. Pac. Fish., 19–22 (1905)
5. Cunningham, G.: Oregon's first salmon canner 'captain' John West. Oregon Hist. Soc. **54**, 240–248 (1953)
6. Taylor, J.E.: Burning the candle at both ends: historicizing overfishing in Oregon's nineteenth-century salmon fisheries. Environ. Hist. Durh. N. C. **4**, 54–79 (1999)
7. Most, S.: Salmon people: crisis and continuity at the mouth of the Klamath. Calif. Hist. **84**, 5–22 (2007)
8. Ralston, H.K.: The 1900 Strike of the Fraser River Sockeye Salmon Fishermen. University of British Columbia (1965)
9. Muszynski, A.: The dialectics of cheap wage labour. In: *Cheap Wage Labour: Race and Gender in the Fisheries of British Columbia*, pp. 129–179. McGill-Queen's University Press (1996)
10. Nash, R.A.: The 'China Gang' in the Alaska Pacific Association Canneries, 1892–1935. In: *The Life, Influence and the Role of the Chinese in the United States, 1776–1960*, pp. 257–284. Chinese Historical Society of America (1976)
11. Kipling, R.: From Sea to Sea: Letter of Travel, vol. 1. Doubleday & McClure Co. (1899)
12. O'Bannon, P.W.: Waves of change: mechanization in the Pacific Coast canned-salmon industry, 1864–1914. Technol. Cult. **28**, 261–287 (1987)
13. O'Bannon, P.W.: Technological change in the Pacific Coast canned salmon industry 1900–1925: a case study. Agric. Hist., 151, 4775–166 (1982)
14. Friday, C.: Organizing Asian American Labor: The Pacific Coast Canned-Salmon Industry, 1870–1942. Temple University Press (1994)
15. Muszynski, A.: Marx's labour theory of value a critique. In: *Cheap Wage Labour: Race and Gender in the Fisheries of British Columbia*, pp. 24–50. McGill-Queen's University Press (1996)
16. Muszynski, A.: Race and gender: structural determinants in the formation of British Columbia's salmon cannery labour forces. Can. J. Sociol./Cah. Can. Sociol. **13**, 103–120 (1988)
17. Stacey, D.A.: Sockeye and Tinplate: Technological Change in the Fraser River Canning Industry, 1871–1912. British Columbia Provincial Museum (1982)
18. Beach, R.: The Silver Horde. A.L. Burt and Company (1911)
19. Ralston, H.K.: Patterns of trade and investment on the Pacific Coast, 1867–1892: the case of the British Columbia salmon canning industry. In: Ralston, J.F., Friesen, B. (eds.) Historical Essays on British Columbia, pp. 37–45. Gage Publishing Ltd. (1980). https://doi.org/10.14288/bcs.v0i1.582.g625
20. Newell, D.: The rationality of mechanization in the Pacific salmon-canning industry before the second world war. Bus. Hist. Rev. **62**, 626–655 (1988)
21. Muszynski, A.: Cheap Wage Labour: Race and Gender in the Fisheries of British Columbia. McGill-Queen's University Press (1996)
22. Irving, W.: Astoria. Baudrey's European Library (1837)
23. Collins, J.H.: Canned Salmon – the hurry-up product. In: *The Story of Canned Foods*, pp. 137–158. E.P. Dutton and Company (1924)
24. Anonymous: Chinks disliked machine. Pac. Fisherman. **23** (1905)

Chapter 6
Canning Soup in New Jersey

Abstract Camden, New Jersey 1862-1935. The development of the Campbell's soup company under the leadership of Jack Dorrance. Managing technology and labor for both large scale production of complex recipes as well as the annual tomato glut. Labor unions and management gurus. Canners must work with their consumers to make brands people will choose to buy.

6.1 Tomatoes in the Garden State[1]

Tomatoes were first domesticated in South America and introduced to Europe after the Columbian Exchange. Over the centuries, they became established as part of European cuisines, and it was European settlers who first brought the plant to North America. The first records of tomato growing in New Jersey date only from about 1812, but, by the 1830s, the southern part of the state was known as a center of cultivation. The fine summer climate and improvements in transportation made the Garden State the ideal foodshed for the growing cities of New York and Philadelphia.

However, the profit for commercial growers was often to be found early in the season when the fruit were less plentiful and people would pay a premium. When the late summer glut finally arrived, tomatoes became so cheap it often made more economic sense to simply plow the crop back into the ground and try again the following year.

Canning offered an obvious way of dealing with the excess, and the tomato canneries opened in New Jersey were some of the first in the country. One of the very first was established in 1847 by the groundskeeper of Lafayette College in Easton, Pennsylvania, on the New Jersey border who packed tomato sauce in tin "pails" with soldered lids and sold them in New York City. Around the same time, a utopian socialist commune known as the North American Phalanx set up a canning factory in Monmouth County in the center of the state, where members of the Bucklin

[1] The story of nineteenth-century farming and canning in New Jersey is taken from histories by Sim [10] and Smith [4].

J. Coupland, *Containing Nature*, https://doi.org/10.1007/978-3-032-11882-0_6

family packed fruit and vegetables and even developed various small machines such as the "Cyclone" pulping machine and a filler capable of filling several cans at a time.

In the next decade, many seasonal canneries were opened by local entrepreneurs looking to take advantage of the tomato glut. In a reinforcing cycle, tomato cultivation thrived across southern New Jersey in response to the growing demand from these canneries. Seeds were planted out in beds in the second week of April, as soon as the fear of frost had passed, and then the seedlings moved to the prepared fields in late May. By summer, the yellow flowers and the smell of the vines dominated the landscape. Picking began in the heat of mid-August and ran until the first frost, usually at the start of October. Throughout the harvest, carts of tomatoes filled the roads as the canneries strained to pack the crop before the fruit spoiled.

Before 1860, the canneries were quite small and local, but federal contracts during the Civil War allowed the development of much larger plants. For example, a plant in Lambertville on the Pennsylvania border was reported to have had a staff of 20 people in 1863 to pack 50,000 cans of tomatoes over the course of the season, but, by 1884, this number had risen to 340,000 cans and by 1891 to half a million. The largest of the canneries was just across the Delaware River from Philadelphia, in Camden.

6.2 Foundations of the Campbell Soup Company

As early as the 1830s, passengers could ride the train from New York to Camden and then transfer to a ferry across the Delaware into Philadelphia. Camden's transportation connections continued to be a source of prosperity and the area rapidly industrialized.

When Abraham Anderson, a local tinsmith, wanted to open a cannery, he chose Camden because it was close to the local farms and had good transportation and an abundant local workforce. By 1868, he was packing 50,000 cans a year of a variety of foods, mainly tomatoes, as well as other pickles and preserves. The following year, Joseph Campbell, a produce wholesaler, joined the firm as a partner and their business grew. They packed a range of foods, but their most famous was canned beefsteak tomatoes whose label showed a single enormous tomato carried on a pole between two men straining under its vast weight. They even won an award for their canned foods at the 1876 Philadelphia Centennial Exposition.

Campbell eventually bought out Anderson's stake in 1876, trading initially as Joseph Campbell & Company, and brought in other investors, including Arthur Dorrance. Campbell resigned in 1894, and Dorrance became president. His family would go on to dominate the company for the next century.[2] When Arthur Dorrance took over the company, it was known for canned tomatoes as well as other canned

[2]The company changed name several times in the following years before settling on the Campbell Soup Company in 1922. To avoid confusion, I will refer to the organization as Campbell's for the remainder of this chapter.

vegetables, ketchup, and a range of other pickles and preserves. In 1897, Arthur Dorrance hired his nephew John T. (Jack) Dorrance, and Jack was only interested in soup.

Jack Dorrance was born in 1873 in Bristol, Pennsylvania, 30 miles up the Delaware River from Camden. His family was rich. His grandfather had made the family fortune with transportation, lumber, and flour businesses, and his uncle, Arthur, invested that capital in high-growth companies like Campbell's. Jack earned a degree in chemistry from MIT and a PhD in organic chemistry from the University of Göttingen.[3] He turned down faculty positions at Cornell and at Columbia University before he was hired by his uncle. The approved company history notes John's low starting salary ($7.50 per week) and the fact he had to pay to outfit his own laboratory, but whether these financial inconveniences made any real economic impact on the wealthy young man is unclear [1]. Jack Dorrance became a director in 1900, general manager in 1910, and, following his uncle's retirement in 1914, president. In 1915, he bought up all the available stock in the company, including his uncle's share, and became the sole owner. When he died of a heart attack in 1930, his estate ($120,000,000) was the third largest the country had ever seen.

One of Jack Dorrance's brothers described him as being "in love with soup" and his passion carried over to his leadership of the company.[4] Some writers claim Dorrance brought his love for soup back from Germany, where it played a more important role in the diet. This may be true, but Americans already knew soup and many regional varieties (e.g., Louisiana gumbo, New England clam chowder) were widely enjoyed and already available in cans. Indeed, Campbell's had been selling canned soup since 1895. Dorrance is also sometimes even credited with inventing condensed soup himself, but as he only started at the company late in the summer of 1897, there would not have been time for him to contribute significantly to a product launched later that very year [2]. Nevertheless, throughout his career, Dorrance focused on making better condensed soup.[5]

By 1910, Dorrance had established a menu of 21 varieties of soup (asparagus, beef, bouillon, celery, chicken, chicken gumbo, clam bouillon, clam chowder, consommé, julienne, mock turtle, mutton broth, oxtail, pea, pepper pot, printanier,

---

[3] As an MIT undergraduate, Dorrance would have been classmates with the pioneer of canning science, Sam Prescott (Chap. 8). The company did not engage in significant corporate philanthropy in the early twentieth century, but in 1950, they made a donation to MIT to support the John Thompson Dorrance Laboratory of Biology and Food Technology (later named Building #16) at the Massachusetts Institute of Technology [11]. MIT Dean emeritus Prescott spoke at the opening symposium for the new building.

[4] *Fortune* magazine captured this reverential attitude to their product in a 1935 profile of the company: "Their attitude toward soup may be judged by the fact that in inter-office memoranda the word is more often than not spelled with a capital S. 'We consider,' says one of them grimly, 'that we have a Soup Service to render to America'" [3].

[5] The one exception to this rule was canned pork and beans, and this product only made sense due to the logistics of the factory. The plant would be closed on Sunday, and, because it took a long time to boil the bones to make stock, the production lines were free to make something that didn't require stock on Mondays. That something was pork and beans.

tomato, tomato-okra, vegetable, and vermicelli-tomato), all packed in 10.5 ounce cans, and all advertised for sale at 10 cents. The labels featured the medal they won for their products at the 1900 Paris Exposition. The red-and-white colors were chosen by a manager who thought the Cornell football team looked particularly smart in their 1897 game against the University of Pennsylvania. Eighteen years before Andy Warhol was born, Campbell's cans had their iconic look. All Dorrance had to do was make them in quantity and sell them at a profit.

6.3 Managing Technology

In some ways, the Campbell's manufacturing operation was simple—they brought in ingredients, made soup, filled it into cans, cooked the cans, and shipped out the finished product. However, there were essentially two different businesses running in the same factory, and each had its own needs.

First there were the 20 varieties of soup they made year-round. These required a staggering range and volume of raw materials that had to be sourced, preserved, shipped, prepared, and incorporated into soup in different ways. In 1935, *Fortune* magazine estimated Campbell's used 116 different ingredients, including beef for stock (they were estimated to use the equivalent of 1440 cows per day), pork (8400 pigs per day), oxtails, calf heads (used in mock turtle soup), sherry (again used in mock turtle soup; the Camden plant had its own wine cellar), leeks, carrots, noodles, rice, chicken, clams, okra, onions, rice, barley, and spices. They bought in so many canned ingredients for some soups that they made a business melting down the empty cans and selling the metal as counterweights for sash windows [3].

The second business run by Campbell's in their Camden factory was tomato soup. For 2 months in late summer, the South Jersey tomato glut arrived at the factory in an endless stream. Perhaps 300,000 baskets of tomatoes had to be processed into soup each day before they began to rot and before the next day's delivery arrived. The operation of the plant was transformed to meet the rush. For most of the year, the factory was coping with the complex supply chain issues required to make sure retailers were well always stocked with all Campbell's varieties—production had to be flexible. However, in the late summer, their worries shifted to maximizing capacity to make the year's tomato soup—production had to be fast and simple.

In 1905, the vast Campbell's factory that did these two jobs had an ammonia compressor for refrigeration, its own generators for electrical power, and a steam boiler burning 20 tons of coal a day. The factory was tall rather than wide; raw material entered at the top and made its way down by gravity power through each stage of processing. Finished cans were labeled on the ground floor and then shipped out to warehouses, delivery trucks, railcars, and barges. Known as "Plant #1," it eventually expanded to cover three blocks of interconnected buildings along Front Street in Camden. While Campbell's never fully succeeded in designing a single building optimized for both core operations, they made the system work by adapting their technology and supplementing the workforce as needed.

Raw ingredients were received, graded, cleaned, and prepared for canning. Because of the wide variety of different operations required for each ingredient, there was initially little automation of these processes. Still, the tasks required were clearly delineated and workers were assigned to specific duties. One particularly unpleasant example involved the preparation of calf heads for mock turtle soup, which required workers to perform specific tasks: skinning the faces, splitting the skulls, removing the teeth, and scraping the nostrils.

The soup itself—the liquid portion of the product—was cooked in rows of large steam-jacketed kettles raised on platforms above the production floor. Each kettle was operated by a worker who stirred the soup with a massive wooden oar for about 30 min. Once the soup was ready, the operator opened a valve at the bottom, draining the contents into a "soup car" that could be wheeled to the canning line. The kettle was then quickly washed and prepared for the next batch.

This process was only partly automated in the 1930s with mechanical stirrers, but even then the need for flexible production meant the soup had to be prepared and taken to the canning line in batches [4]. The solid ingredients, the "garnish," were weighed into the cans by hand, then the liquid soup was added using a filling machine, and the cans were closed, initially using soldering machines, but later with solder-free double seams. The retorts were loaded using an electrically powered crane and cooked at 100–127 °C (212–260 °F), for anywhere from 20 min to over an hour. The cans were cooled, their iconic red-and-white labels were applied, and they were ready to be shipped.

During the tomato season, the plant converted to a different mode of production. Huge lines of carts, and later trucks, crowded the streets of Camden and formed long lines at the gates before dawn. The loading docks opened at 5:45 AM and closed at 6 PM, meaning the factory ran until midnight or later. An inspector graded the tomatoes to determine the price paid: #1 were perfect, while #2 had blemishes and defects. The tomatoes were washed, and then women worked quickly with knives to cut out any defects.

Next, the tomatoes were broken apart and cooked, and then the seeds and skin were separated in a centrifugal machine and discarded. The crushed, cooked tomatoes were transferred to soup kettles where they were combined with butter, salt, flour, sugar, onions, and spices. For most other soups, batches were wheeled from the kettles to the canning lines in soup cars. However, because tomato soup required no additional solid ingredients and was produced in much higher volumes, it could be piped directly from the kettles to the can filler [3].

Indeed, the volume of tomato soup production was so high that Campbell's lacked the warehouse space to store all the finished cans. To solve this, the company offered steep discounts on purchases made during tomato canning season, effectively compelling wholesalers to buy their entire annual supply at that time and store the cans in their own warehouses.

Aside from the food, the other thing a large cannery needs in quantity is the cans themselves. For most of the time John Dorrance was in charge, Campbell's bought so many cans from the Continental Can Company that Continental built a dedicated can factory next door to the cannery [5, 6].

The Continental Can Company was formed after a failed attempt by two brothers, William and James Moore, to create a monopoly, a "trust" in the language of the time in can manufacturing at the start of the century. The Moores had become wealthy by merging companies across a range of industries to achieve efficiencies and a stronger market position. Notably, they were involved in the tram that consolidated a group of midwestern bakers in 1898 to form the National Biscuit Company—Nabisco. In 1901, they turned their attention to can making.

They partnered with the experienced can maker Edwin Norton (see Chap. 4 for his role in establishing the first truly integrated can factory) to negotiate purchase options with over a hundred individual can-making companies. These companies were then consolidated into the American Can Company, which came to control 90% of the market. The Moores used this monopoly power to raise can prices, but new competitors quickly emerged to undercut them.

Surprisingly, the most successful of the new competitors was led by the same Edwin Norton who had stepped down from his role at the Moore brothers' American Can Company, claiming poor health. He miraculously recovered and immediately went back into business as the Continental Can Company. By 1935, Campbell's had become Continental's largest customer, accounting for about a third of its total sales. However, the relationship ended abruptly in 1936, when New Deal legislation prohibited bulk discounts. Campbell's were unwilling to pay the same price as a customer "buying a single carload," so they simply set up their own factory to make their own cans.

6.4 Managing Labor[6]

Despite efforts to mechanize, making soup required many human hands. The population of Camden when the factory first opened were German, British, and Irish immigrants, but by the 1920s, it had shifted to largely Italian and Eastern European. These were the people who provided the workforce for the cannery.

In 1896, the Camden plant had a staff of about 25 year-round and 300 during the tomato canning rush, but by 1935 this had risen to 2500 with double that during tomato season [3]. The dependency on seasonal labor led to a "two tier" labor market with the temporary summer workers drawn from wherever the company could find people—migrants, agricultural laborers, high school students, and housewives. Most of the temporary workers were laid off at the end of the season, but a few of the best performers were recruited into the permanent workforce. Over half of the workers were women, and they were paid less than men, with "boys" rated somewhere in between.

[6]The history of labor in Campbell's plants is taken from Daniel Sidorick's work [7–9] as well as the *Fortune* magazine articles from 1934 [5], 1935 [3], and 1955 [11].

The work was tough. New Jersey is hot in the summer, and the heat next to the steam kettles must have been unbearable. One worker recalled having to "peel" her clothes off at the end of a shift. Some tasks were hazardous. Besides the "normal" risks of sharp knives, hot steam, and heavy lifting, there were particular risks to canning. For example, one of the first jobs offered to "boys" was loading and unloading filled cans into the retort baskets. The company provided gloves to protect their hands from the hot cans, but these were inadequate, so they often wrapped their fingers in cardboard sandwiched between two pairs of gloves [7]. The factory was also loud, with metal cans banging against metal equipment. Some workers later sued the company for the hearing loss they suffered as a result.

The work week at the start of the century was 53 h, which was not untypical for the era. However, during the tomato harvest, Campbell's expected 12 h a day, 6 days a week, and 5 h on Sundays. In 1933, the work week was lowered to 35 h in response to the unemployment caused by the Great Depression. There was an hourly pay raise associated with the reduction, but workers still made less per week. Other New Deal legislation had supported labor organizations, and Frank Manning, the socialist president of the Unemployed Union of New Jersey, tried to unionize the Campbell workforce in response to the wage cut. On April 1, 1933, the union organized a strike and closed the factory down, demanding better pay and conditions. The company fought back, even proposing a "company union," the Campbell Employees' Representation Plan, to give the impression they were responsive to worker concerns. There were some violence, marches, and many arrests. At one point, a group of strikers surrounded Camden City Hall demanding the release of prisoners. In the end though, the union agreed to return to work on May 5th with a 7% pay rise but little else.

Campbell's resisted the demands of the union because controlling costs, particularly labor costs, was essential to maintaining profitability. Their only product was prominently advertised for retail sale at ten cents, and this price became so fixed in the minds of Campbell's executives that for decades they went to extraordinary lengths to cut costs and avoid a price hike. Margins were tight, and initially Campbell's made only $2.50 on a thousand cans, but this had fallen to only $1.67 by 1923 [4]. Every operation, every transaction, was scrutinized to defend that tiny margin. The workforce had to be organized alongside the machines and the ingredients into the most efficient system possible. In terms of labor relations, "efficiency" meant paying as little as possible for the work needed. It is not surprising there was conflict. However, when conflict did arise, it was often not only about total wages but also the ways in which the company decided how much a job was worth.

Powered machines form the ideal of industrial efficiency, by repeating the same task over and over again in a predictable way. When a factory is working perfectly, the labor of humans fits into the same mechanical rhythm; but people are not machines—they have their own priorities. The job of the production manager is to get the machines and people to work effectively together. Campbell's had initially used a familiar combination of carrots (in the form of production bonuses) and sticks (bullying foremen and threats of firing) to motivate their workforce. However, the growing complexity of his business convinced Jack Dorrance that he needed a more modern approach to management.

In the early years of the twentieth century, many industries were faced with similar challenges of labor and machines, and the concept of "scientific management" emerged as a solution. The most prominent advocate was industrial engineer Frederick Winslow Taylor, and the entire approach came to be known as "Taylorism." Taylor closely analyzed individual tasks in industrial processes—such as loading a barrow, plucking a chicken, or soldering a seam—and identified the most efficient method for each. He documented these methods and insisted that workers be trained to follow them precisely, not according to personal preference but according to what he had determined was best.

This approach allowed managers to better plan labor needs and evaluate worker performance. Any underperforming employee could be "fixed" by retraining or replaced much like a malfunctioning piece of equipment. Taylorism inspired a generation of managers seeking to rationalize labor in industrial production—and also a wave of imitators promising quick efficiency gains. Among them was the colorful Mr. Charles E. Bedaux.

Charles Bedaux[7] was born in France and moved to the United States in 1906. He worked in a variety of menial jobs, moving between America and France, before taking a position as an interpreter for an Italian consultant who was visiting the United States to study Taylorism. Bedaux developed this minimal experience into his own system, which became popular in Britain and the United States in the interwar period and made him rich. Bedaux argued, uncontroversially, that any task a worker could perform required a certain period of effort and a certain period of recovery. More dubiously, however, he claimed to have discovered a scientific formula to work out the appropriate recovery time, and, using this, he could assign a Bedaux unit of time to the task (B): the time to do the work and the time to recover. Workers were expected to produce 60 B per hour regardless of the work and got a bonus if they went faster. On the other hand, workers achieving less than 60 B were at risk of dismissal.

The results of the Bedaux system appeared to be as impartial and authoritative as mathematics. However, the basis of his calculation of recovery time was obscure, and Bedaux never properly explained it. Indeed, it may not have even existed. The labor historian Daniel Sidorick argued that Bedaux "dispensed with any efforts to

[7] Bedaux's remarkable life is described in three articles by Janet Flanner [12–14] published in *The New Yorker* in 1945, his economics by Levant and Nikitin [8], and his effect on labor at Campbell's in Sidorick's book [7]. Scenes from Bedaux's biography illustrate to his boundless self-confidence and endless self-promotion. He led pioneering automobile expeditions across Canada, Asia, and Africa in the 1930s. He hosted the wedding of Wallis Simpson to the Duke of Windsor, the abdicated Edward VII of England, at his French chateau and soon afterward arranged their notorious tour of Nazi Germany. During the war, he worked with Vichy government and collaborated with the Nazi invaders. One notable wartime project was an effort to organize the government of a small town, Roqueforte des Landes, on the utopian idea of "equivalism" where his B units served both as wages and as a currency. He was arrested by Allied forces in Algeria late in the war where he had been planning to build a trans-Saharan railroad and a parallel peanut-oil pipeline for the Vichy government (Bedaux's life is full of bizarre details). He died in Miami, apparently by suicide, while awaiting trial for treason and for trading with the enemy.

understand and redesign the work process, and got down quickly to the bottom line: lowering costs by speeding up labor, though the whole exercise was, to be sure, wrapped in scientific trappings" [7]. Nevertheless, Dorrance was impressed by the quick improvements in productivity promised and adopted the Bedaux system at Campbell's in 1927.

Every task in the cannery was studied by Campbell's industrial engineers with the help of Bedaux consultants and assigned a B value. For example, cooking a basketful of tomato soup in a retort was worth 5.1 B, and a basket of oxtail soup was worth 7.8 B; therefore, each worker could be expected to cook 11.7 and 7.7 baskets per hour, respectively. The grisly tasks of face skinning, skull splitting, teeth removal, and nostril scraping necessary for mock turtle soup preparation were rated as 2.60, 2.30, 2.29, and 2.36 B per calf head [8].

The Bedaux system was appealing to Campbell's management, as it reduced the complex operations of the plant to a single number they could seek to optimize. The appropriate amount of labor needed for a desired level of productivity could be calculated. Bedaux charts could be used to compare workers in different parts of the plant and reward or punish workers, supervisors, and managers accordingly.

The workers on the other hand hated the system. They found it arbitrary and hard to understand. What exactly justified a B rating? The system might say that peeling and chopping a basket of onions takes a certain amount of time, but this assumed that an onion was a standard thing. If, on the day you were working, you were faced with small or misshapen onions, you would struggle to complete your assigned work and be penalized through no fault of your own. Furthermore, tasks could be reassessed at the request of management in ways inevitably left workers being expected to do more work in less time. Daniel Sidorick illustrates this point with an example: "Before Bedaux, twelve men loaded 28,000 cans onto the labeling machine lines; afterward, just five had to load 31,500 cans, with the additional assistance of only the Bedaux incentive system" [7]. Workers at Campbell's called the Bedaux system the "speedup" system. Anger at the apparently irrational demands from management justified by the Bedaux system was part of the union grievances which lead to the 1934 strike. However, while management was willing to move a little on wages, they were not willing to give up the control the system offered.

Campbell's Plant #1 was located where it was because, in the 1860s, Abraham Anderson wanted to be close to both the abundant tomato harvests from southern New Jersey farms and the labor force provided by Camden's growing population. While other Camden employers eventually moved out in search of cheaper or less militant workers, the tomatoes still had to be processed near where they were grown. Campbell's were forced to stay and fight to retain a workforce at rates they felt justified by tight profit margins. Only much later did they move much of their production to more modern facilities in other states. Later still, the use of California tomato pulp to make soup finally severed the company's historic dependence on Jersey tomatoes and Camden workers. Production of soup in Camden stopped in 1980, and, in 1991, Plant #1 was finally demolished.

6.5 Managing Brands

The need to make people, food, and machines work closely together is common to all canneries, and Jack Dorrance managed it effectively at a large scale during his leadership at Campbell's. Dorrance, however, faced additional challenges because Campbell's soup was a complete meal, not just an ingredient. Their cans were simply opened, diluted, heated, and eaten—and they had to taste good every time. Their goal was therefore not simply to preserve the qualities of a natural product like salmon; instead, they had to combine mixtures of ingredients to make recognizable varieties of delicious soup.

The actual processes of product development in the early years of the Campbell's company are not well documented. Company legend emphasizes the scientific qualifications of their leader. Certainly, Dorrance was proud of his education and insisted on using his "Doctor" title throughout his career. One of his first acts when he was hired had been to establish a laboratory in the factory; however, it is not clear what he used the laboratory for. Photographs of his laboratory show typical glassware and reagents, yet it was not obviously equipped with the kitchen facilities to make soup at a small scale. Food chemists in his time knew how to measure the water, fat, and protein content of foods, all of which would have been useful to ensure product quality and consistency, but not necessarily for developing better varieties of soup.

Dorrance recognized this and is quoted as saying: "Chemical analysis is valuable but it only goes so far. It may determine the nutritive value, but in the marketing of any successful food product you will notice that these are the prime considerations that have played the most important part of their success with the public: first, appearance – it must please the eye, second, odor, third taste or flavor – it must please the palate. You may pack a food product full of vitamins, calories and those necessary units of nutriment, but unless appearance, odor, and flavor are right, it will never sell" [1]. The fact that his predecessors had been making soup before he joined the company meant that whatever facilities they used were already established although not remarkable enough to be described as a "laboratory." There were already skills at work at Campbell's beyond the science Jack Dorrance had learned in college.

In addition to his pride in being a scientist, Jack Dorrance was proud of his culinary skills. He claimed to spend 3 months a year refining his craft by working as a chef in some of the best restaurants in London, Paris, and New York and then, at the end of his shift, changing and returning to the same restaurant for his own dinner. He even boasted that he had made soup for the Prince of Wales.[8] Regardless of the truth of these stories, Dorrance certainly recognized that the creativity of a chef was just as useful as that of a scientist. In 1902, Campbell's hired Louis Charles de Lisle

[8] This anecdote is based on a quotation from Dorrance given in the approved company history [1]. However, the source of the quotation is not given, and it seems remarkable that someone with Dorrance's responsibilities could regularly find time for this sort of experience, however valuable it might be.

as an executive chef. Lisle had been born in France but emigrated to America where he worked first as a chef and then for Campbell's competitor, the Franco-American Food Company. It seems likely that men like Dorrance, and later Lisle and his successors, adapted soup recipes used in domestic and commercial kitchens to make them work in the more stringent conditions of the steam kettle and the retort.

A second challenge faced by the Campbell's chefs translating domestic recipes to the factory was consistency. The chef in a restaurant tries to make the best soup possible from the ingredients available that day. By contrast, every can of Campbell's soup had to taste the same, regardless of any variability in the ingredients. Product consistency was important because Campbell's soup was a branded product. The iconic red-and-white labels signaled to the consumers the contents were what they expected, and Campbell's had to deliver on that promise.

Even when the recipe had been worked out, managers had to ensure it was implemented consistently in the factory by workers who might not otherwise share their bosses' passion for soup. Product quality and consistency were deeply rooted in corporate culture. Anderson and Campbell had started the company tradition of a "tasting ceremony" when company senior leadership would gather to sample and discuss the latest product. The tradition continued throughout Dorrance's leadership and, in 1935, *Fortune* magazine described Arthur Dorrance, Jack's younger brother and successor, leading the ceremony as if it were a religious rite: "Every morning at eleven o'clock the Campbell management drops its other tasks and, headed by President Arthur Dorrance, assembles in the experimental kitchens to taste the day's run of soup. …they stand about, spooning mouthfuls of mulligatawny or pepper pot or chicken, and pronounce it good. Then they go back to their desks and at twelve-thirty the six members of the No. 1 team meet again in the executive's dining room, its walls lined with framed proofs of Campbell advertising, for a lunch which begins with soup and which is, in a large measure, devoted to talking about soup. There is usually something of a hush here, for the Campbell management is made up, by and large, of what Sinclair Lewis once called Men of Measured Merriment—and a belly laugh might topple something over" [3]. The fact that senior leadership were so very seriously invested in product quality is a good indication that similar care was taken in the lower rungs of the company.

Another way Campbell's were able to maintain the quality and consistency of their products was by controlling and standardizing more of their supply chain. All goods received were checked carefully for quality, and any substandard items were rejected or repurposed. For example, discolored barley grains were sold to the US Army Signal Corps as feed for messenger pigeons. As the company grew, they became more able to insist that suppliers delivered exactly what they wanted to buy. In 1909, Campbell's started to develop a tomato breeding program on the grounds of Dorrance's home in Cinnaminson, New Jersey [4]. They hired a plant breeder, a professor from the University of New Hampshire, and collaborated with the Experiment Station of the local state university to develop the "Rutgers tomato" which offered just the right balance of sugar and acidity, the right shape, and the right color. They provided local farmers with seeds to ensure they got what they wanted and taught them ways to improve the productivity of their farms.

Alongside making a recognizable and desirable product, Campbell's worked hard to ensure consumers understood just how good their soups were. They wanted people not to simply ask their grocers for soup but to demand it be Campbell's. This was a different strategy to that pursued by their rivals Heinz, who used a large sales-force to convince the retailers to stock their products. While Campbell's direct-to-consumer approach required a smaller sales staff, it required much more advertising [9]. Their advertising budget grew from $50,000 in 1901 to $135,000 in 1904, $500,000 in 1912, and, in 1920, one million dollars.

Campbell's recognized early on that the people who selected their products were women, usually busy working women on a budget, who were looking to feed their families. With this target market in mind, their messages consistently conveyed the values of convenience, quality, and value. Some of the initial adverts were printed cards in public transportation because, as the company secretary, Leonard Frailey, put it: "we believe we can reach the clientele we are most after, that is, the women. They are frequent riders on streetcars." Later they became pioneers of print ads in women's magazine such as *Good Housekeeping* where they established what came to be known as the "Campbell's Soup position"—a full page ad to the right of a page of solid text. In 1904, they hired the artist Grace Drayton to develop the "Campbell's kids" as mascots for the brand. The "kids"—anime-like toddlers with huge heads and eyes—were featured in many of the print adverts and later appeared as collectable dolls and animated in TV commercials.

Beyond direct advertising, Campbell's used other media to teach women how they could live happier and more fulfilled if only served more soup. In 1910, they printed a book of sample menus, unsurprisingly featuring a lot of soup courses, aimed at anxious housewives worried about entertaining guests in a socially acceptable manner. In 1914, they published a recipe book featuring Campbell's soup as an ingredient in other foods. Later still, Campbell's began advertising directly to children through sponsorship of popular TV programs such as "Lassie." In his 38 years in control, Dorrance spent $54,000,000 in advertising, and Campbell's soup was advertised more than any other manufactured product [2].

6.6 Reflections for Modern Technologists: Networks of Experts

In Chap. 4, we used the story of the development of can-making machines to show how industrialization involves the progressive replacement of skilled craft labor by machines. We complicated that narrative in Chap. 5, by using the story of the salmon canneries to show that, while the process seems inevitable in retrospect, at each stage it involves making judgements between using machines and labor. Still, the salmon canneries were isolated operations making a single, simple product.

Campbell's also made canned food with essentially the same types of machines, but their business was vastly more complicated. The differences between Campbell's and the Pacific canneries illustrate some of the ways that canning technology must be successfully integrated with other aspects of food manufacturing:

- Campbell's was much larger and better capitalized than the salmon canneries. They were able to hire more forms of specialized expertise in the form of industrial engineers, horticulturalists, chefs, accountants, management consultants, and marketers. They had complex hierarchies and bureaucracy. The salmon canneries on the other hand were smaller and more entrepreneurial. They had their own racial hierarchies, but a handful of people made the decisions.
- Campbell's was fixed in Camden by their history and previous investments in Plant #1. Campbell's eventually build more efficient plants elsewhere, but they were committed to Camden long after other factories had relocated. Salmon canneries were much more temporary structures, quickly abandoned when pollution or overfishing made a river unproductive.
- Campbell's depended on the population of Camden to provide a workforce. The workers had some local political power and were able to organize themselves in a union to protect their interests. Campbell's responded by calling on management "experts" like Bedaux to try to exert control. In contrast, the salmon canners had to work with the labor contractors to find a Chinese crew, but simple labor scarcities gave the workers some negotiating power.
- Campbell's worked to a regular, predictable annual cycle. They signed contracts with farmers to ensure delivery of the tomatoes they wanted when they wanted them and established a can-making factory on site to exclusively supply their needs. Planning and scale enabled them to deal more efficiently by-products: the barley grains sold for pigeon feed and the vegetable cans melted down to make weights. Careful planning was impossible in the small salmon canneries where the exact timing and size of the salmon run was unknown, and the remoteness of the canneries meant all the materials and workers had to be in place waiting for the fish to arrive. Any excess fish was simply discarded along with the inedible parts.
- Campbell's had a close connection to their consumers through branding and advertising. At the plant level, product development (making a product people liked) and quality control (making it the same year in and year out regardless of the variability in the ingredients) became a much bigger issue. Salmon canners also had to broadly meet their consumer's expectations, but their products were much closer to a simple commodities.

Taken together, the Campbell's story reveals ways food scientists today must fit into a broader array of technical expertise. People who might not know much about how food was canned still had important knowledge to contribute. Successful modern technical work requires this sort of team expertise where individual contributors are strong in their areas but also understand what others can contribute so they know when to listen and learn.

References

1. Morton, R.: America's Favorite Food. Harry N. Abrams, Inc. (1994)
2. May, E.C.: The Canning Clan. Macmillan Publishers Limited (1937)
3. Anonymous: Campbell's Soup. Fortune, 68–76, 124, 126, 128, 130 (1935)

4. Smith, A.F.: Souper Tomatoes. The Story of America's Favorite Food. Rutgers University Press (2000)
5. Anonymous: Profits in cans. Fortune, 76–85, 134, 136, 138 (1934)
6. Pearson, G.S.: The Democratization of Food: Tin Cans and the Growth of the American Food Processing Industry, 1810–1940. Lehigh University (2016)
7. Sidorick, D.: Making Campbell's Soup. In: *Condensed Capitalism: Campbell Soup and the Pursuit of Cheap Production in the Twentieth Century*, pp. 13–41. Cornell University Press (1991)
8. Sidorick, D.: Bedaux, discipline, and radical unions. In: *Condensed Capitalism: Campbell Soup and the Pursuit of Cheap Production in the Twentieth Century*, pp. 42–67. Cornell University Press (1991)
9. Sidorick, D.: Introduction: global strategies, hometown factories. In: *Condensed Capitalism: Campbell Soup and the Pursuit of Cheap Production in the Twentieth Century*, pp. 1–12. Cornell University Press (2017)
10. Sim, M.E.: History of Commercial Canning in New Jersey. History and Early Development. New Jersey Agricultural Society (1952)
11. Anonymous: Campbell takes the lid off. Campbell's Soup, 78–83, 120, 124, 126, 129 (1955)
12. Flanner, J.: Annals of Collaboration; Equivalism—I. The New Yorker. **28** (1945)
13. Flanner, J.: Annals of Collaboration; Equivalism—II. The New Yorker. **32** (1948)
14. Flanner, J.: Annals of Collaboration; Equivalism—III. The New Yorker. **32** (1945)

Chapter 7
The Early Science of Canning

Abstract Europe from about 1665 to 1870. The scientific ideas that will explain canning develop largely independently from the technology of food preservation. An incorrect chemical explanation of why canning works is eventually superseded by a better microbiological explanation – but without making any difference to the practice. How science changes its mind.

7.1 The Science and Technology of Food Preservation

The earliest humans were skilled at identifying the plants and animals in their environment that they depended on for survival. They knew how to prepare them as foods, making them safe and enjoyable to eat, and they often shared food within their groups. As people began to settle into more sedentary agricultural societies following the Neolithic revolution, the need for food preservation grew, involving methods such as salting, smoking, fermenting, and drying. Larger groups provided more scope for specialization of skills, and some people became experts at food preservation and others learned to write.

Some of the earliest examples of written language are concerned with economic matters such as harvests, trade, and taxation. However, while the farmers and artisans who prepared and preserved food remained largely illiterate, a small, educated elite began to observe and document their practices. The books and letters they wrote provided a new way for distant groups of educated people to share knowledge.

One example of this sort of writing is the *Geoponika*, a tenth-century Byzantine collection of agricultural wisdom [1]. It includes a method to preserve apples by sealing them hermetically: "wrap each apple in dry fig-leaves, then cover them with white potters clay, and lay them up when dried in the sun, and the apples will remain as they were put in." We can imagine a Roman aristocrat reading about this technique and instructing his steward to instruct his slaves to try it on his estate.

This marks the beginning of a distinction between practitioners who knew through hands-on experience and those who knew from a distance, through reading, and often delegated the actual work to others. That distinction forms the basis for

J. Coupland, *Containing Nature*, https://doi.org/10.1007/978-3-032-11882-0_7

thinking about the science of canning as something separate from the technology of canning.

In previous chapters, we met some of the first canning technologists—Appert, Girard, Donkin—who built businesses preserving food in a new way. They were middle-class men who rose to own factories, but for most of their lives, they were immediately concerned with the practical work required. They made their livings from it. In this chapter, we turn to the scientists who became interested in why food preservation worked. The line between science and technology is often blurred—both are simply different forms of human knowledge—but the scientists we will meet here tended to be more removed from day-to-day practicalities and often sought to maintain a degree of distinction in their social status. This chapter focuses on the early science of canning, up to around 1870, when the scientific foundations of Appert's work were becoming widely understood. Later chapters will show scientists interested in canning interacting with technologists in very different ways, as the industry grows and science takes on the role of its servant.

7.2 The Science of Canning Before Canning[1]

Very few of the earliest writings on food preservation offer much insight into why the methods they discussed worked. However, we can infer some ideas from ancient beliefs about the causes of disease. Many ancient peoples believed that disease was due to exposure to "bad air" [2]. According to this view, air could become contaminated with "seeds" of disease by exposure to hot, wet earth. These seeds could then settle onto bodies and spread the sickness.

Food scientist and historian Norman Cowell suggests that similar thinking may have influenced ancient understanding of food preservation. So, for example, sealing a food away from air as the author in the *Geoponika* had proposed might have been seen as a way to protect it from contamination. Similarly, if heat plus moisture made air "bad," then cooling or drying a food would logically help delay spoilage. The idea of heat as a preservative method was not as philosophically well-established in antiquity, presumably because cooked foods spoiled as easily as raw. Practical methods for preserving food by sealing it away from air—often in combination with acids, salt, smoking, or heat—continued to be collected and anthologized throughout the Middle Ages and into the early modern period. When explanations were offered, they were typically framed in terms of the ancient "bad air" theory.

However, in the seventeenth century, some Europeans began to question traditional wisdom and sought explanations for natural phenomena based on direct observation. This shift marked the beginning of the Enlightenment, a period

[1] The science of food preservation before and during Appert's time is largely adapted from Chapters 2, 3, 4, and 5 of Norman Cowell's PhD dissertation [3]. Other sources are cited as appropriate.

characterized by a growing emphasis on empirical evidence and rational inquiry. A key figure in this shift was Robert Boyle, a wealthy Anglo-Irish aristocrat and natural philosopher. Boyle was deeply interested in how the emerging sciences could improve everyday life, and one obvious area of need was better food preservation.

Boyle began his studies the way scholars had done since antiquity, by making an anthology of the methods currently or historically used to preserve food. He was far too wealthy to have preserved his own food, but he gathered methods from books, from correspondence with friends, and from his own observations and described them in his 1663 book, *Some Considerations Touching the Usefulnesse of Experimental Naturall Philosophy* (cited by Cowell [3]). In this, Boyle described biscuits sealed in tin-lined barrels, gooseberries packed in bottles, and cooked meats stored under a layer of fat. He next connected these examples with an explicit scientific theory: "that Air doth much contribute to the corruption of some Bodies, and the exclusion of Air to the hindering it." What Boyle had done up to this point was very much in the long tradition of scholars collecting the practices of working people and explaining them through inherited ideas. But Boyle was also part of a new tradition that sought deeper understanding through practical experimentation.

Boyle made a connection between the practical problem of food preservation and one of his major scientific interests—the properties of gases. In the late 1650s, he built an air pump capable of creating a vacuum inside a glass chamber. By observing how objects behaved in this vacuum, he could infer properties of air itself. For example, he demonstrated that a vacuum could both asphyxiate small animals and extinguish flames, leading him to conclude that both life and combustion depended on certain components of air. This apparatus also allowed Boyle to explore how the presence—or absence—of air might contribute to food spoilage. In doing so, he brought experimental tools to bear on a question that had long been addressed only through tradition and observation.

In a series of experiments in the 1660s and 70s, Boyle showed some foods could be preserved to a degree by storing them in a vacuum. However, his methods were far from perfect, and it seemed in many cases heating in combination with the vacuum worked better. For example, raw oysters stored in a vacuum spoiled, while cooked oysters were stable for 3 weeks. Boyle also cooked some meat in Papin's bone digester (an early pressure cooker; see Chap. 4) and then applied a vacuum and found the results were better still ("I have found with elixation [boiling], the perfecter it is, doth so much the more hinder fermentation."). Clearly his original theory ("that Air doth much contribute to the corruption of some Bodies, and the exclusion of Air to the hindering it") needed to be modified to account for the effects of heat.

Denis Papin, the French natural philosopher who had invented the bone digester while working with Boyle, continued to perform experiments in food preservation during his time as curator of experiments for the Royal Society in the 1680s. Papin packed various raw or cooked foods into glass jars, applied a vacuum, sealed them, and then gently heated the sealed jars. Although the secondary heating was mild and his results inconsistent, he succeeded in preserving some foods for several months.

These experiments came close to establishing the twin technological principles of canning—sealing and effective heat treatment—125 years before Appert.[2]

The modern explanation of Papin's results is that the heat killed most spoilage microorganisms, while sealing prevented recontamination. The vacuum may have helped keep the lid in place during heating and may have inhibited the growth of any remaining aerobic organisms, but it was not the critical factor. At the time, however, spoilage was still seen primarily through the "bad air" theory, with heating a very secondary effect.

Although their explanation was incorrect, Boyle and Papin advanced beyond their predecessors by linking real-world observations with controlled laboratory experiments which they used to develop a theory. They shared their findings through public demonstrations, lectures, and correspondence with other scientists across Europe—communications that were increasingly being collected in scientific journals. When later generations of scientists sought answers, they turned to this growing body of published literature. It was this foundation that shaped the ways scientists responded to Appert's invention.

7.3 The Scientific Reaction to Appert's Achievement

According to Norman Cowell, Nicolas Appert's achievement was the result of craft skill and persistence, rather than any prior engagement with the scientific literature of his time [3]. However, when he sought recognition and reward for his invention from the Sociêté d'Encouragement pour l'Industrie Nationale in 1810, the organization turned to scientists to validate his findings and explain how his method worked.

The validation studies reproduced in *Book for All Households* were conducted by a team of four respected scientists and industrialists. They visited Appert's factory, observed the food being packed, and then opened the same bottles several months later to confirm the contents were still good. This was likely just good "due diligence" and not drawing on any scientific principles; however, it is an illustration of the social hierarchies at the time that scientists were seen as trustworthy and qualified to assess the achievements of a technologist.[3]

---

[2]Although not central to our story, it is interesting to speculate why practical canned food did not develop in the seventeenth century. It may have been because the materials—tinplate, solder, and glass bottles—were not adequately developed in this period or because long sea voyages were not yet important enough to create a viable market. It may also have been that the people who made discoveries had interests other than going into business themselves.

[3]Another example of how scientists helped advance the early technology of canning without directly using science was through their transnational social connections. Even during a time when war and blockades limited trade in Europe, scientists were still exchanging ideas in letters and journals. The chemist Claude Berthollet was involved with both SEIN in Paris and the Royal Society in London and so could provide letters of introduction to both Donkin/Girard and Nicolas Appert when they visited Joseph Banks in 1810 and 1814, respectively [27, 28]. See Chap. 3.

More scientifically interesting, however, is a paper written by 1810 by one of the scientists who had conducted the validation at Appert's factory: Joseph Louis Gay-Lussac [4]. Gay-Lussac was at the center of early nineteenth-century European science [5]. Already in his career he had done pioneering work on the thermal expansion of gases (with Claude Berthollet and Pierre-Simon Laplace), he had worked out the relative proportions of oxygen and hydrogen in water (with Alexander von Humboldt), and he made a record-breaking flight in a hot air balloon to measure how the properties of the atmosphere change with height (with a young Jean-Baptiste Biot). He brought his scientific knowledge and reputation to his investigation of Appert's practical achievement.

Gay-Lussac sealed grape juice in bottles and processed them in boiling water according to Appert's methods [6]. When he opened the bottles a year later, the juice was fresh but soon afterward began to ferment into wine. At one level, this could be seen simply as a replication of Appert's work by a known and respected scientist and hence further validation that it worked. However, because Gay-Lussac was already aware of the current theories in chemistry, he was able to use his results to put forward a general scientific theory about *why* wine ferments. To do so, he built on the idea of "bad air," which had been put into scientific terms by Boyle, but incorporated the latest discoveries that air was a mixture of different gases and one of these, oxygen, could react with other materials.

Gay-Lussac argued that oxygen reacted with the juice to form a soluble "ferment," a compound which initiated fermentation of sugar to alcohol.[4] In his view, the heating process in canning drove oxygen from the food and destroyed any existing ferment, while the sealing prevented more oxygen getting in. We now understand the "ferment" as living yeast cells carried in the air and on the surface of grapes, which can be destroyed by heat. Gay-Lussac, however, interpreted his results in terms more akin to chemical catalysis. Heating the catalytic ferment rendered it inactive—an effect more comparable to the curdling of egg white during cooking than to the destruction of a living organism.

Gay-Lussac's theory proved wrong, but a reasonable one based on the evidence available interpreted through his chemist's view of the world. In many ways the fact that Gay-Lussac was wrong made no real difference to the development of canning and certainly not to Appert. The report from a respected scientist added useful credibility, but the methods developed by technologists already worked and did not change based on Gay-Lussac's theorizing.

[4]As Cowell pointed out, later writers on canning came to misinterpret Gay-Lussac as having argued the oxygen directly reacts with food and causes it to spoil, but his real views were more subtle [3].

7.4 Microbiology Catches Up (But Is Ignored)[5]

Ancient peoples quickly recognized that simply being near a sick person during an epidemic increased the risk of contracting their disease and was therefore to be avoided. We can see this wisdom written into ancient texts, for example, the Bible (Leviticus Chap. 13) gives instructions to isolate people with leprosy and to burn their clothes. A theoretical foundation for such practices was proposed in 1546 by the Genoese scholar Girolamo Fracastoro, who argued that contagion could spread through direct contact, at a distance, or via intermediate agents like clothing or wooden furniture. He speculated that disease was transmitted by "seeds" or "germs" that carried infection from the sick to the healthy. To support his theory, he drew a parallel to how spoilage can spread from one piece of fruit to another by contact.

In hindsight, Fracastoro's idea foreshadowed the modern germ theory of disease (and food spoilage), though he was vague about the nature of these germs and lacked the means to study them and his theory was not universally accepted. When the Great Plague swept through London in 1665, the physician William Boghurst could only speculate it might arise "from the maturation of the ferment of the Faeces of the Earth extracted into the Aire by the heat of the sun, and difflated [dispersed] from place to place by the winds" [7]—the ancient idea of bad air.

At the same time Boghurst was speculating about disease and Boyle and Papin were experimenting with pressure cookers and vacuum pumps, elsewhere in London members of the same circle of natural philosophers were beginning to explore life too small to be seen with the naked eye. Robert Hooke, who had served as the Royal Society's curator of experiments, constructed some of the first practical compound microscopes in the 1660s. In 1665, he published some of what he had seen in the richly illustrated *Micrographia*, which revealed intricate details of insect and plant structures that had previously been unknown. His work drew widespread attention and inspired others to pursue further investigations.

Antonie van Leeuwenhoek was a draper in Delft, Holland, who crafted lenses to inspect the quality of cloth. He mounted his tiny (~3 mm) lenses onto a small metal plate and held it close to his eye to examine samples positioned on the head of a pin. This simple, single-lens microscope avoided some of the optical distortions common in compound microscope and, when used skillfully, could achieve up to 250--fold magnification and resolve objects as small as a micron (a millionth of a meter) [8]. Inspired by Hooke's *Micrographia*, van Leeuwenhoek began sharing his observations with members of the Royal Society in London.

Leeuwenhoek was endlessly curious about the new invisible world and studied almost everything he could lay hands on. In a 1674 letter, he described a study of the mechanisms of taste. He had softened peppercorns in water for a few weeks but then was astonished when he found microscopic eels ("animalcules") swimming in it [9]. Most of these organisms were protozoa, but the smallest of the three types reported ("ten hundred thousand of them could not equal the dimensions of a grain of such

[5]The early history of microbiology is taken from Barnett [6] and Bolloch [29].

coarse Sand") were probably the first bacteria ever seen. Hooke copied Leeuwenhoek's methods and confirmed his findings to the scientific network of the Royal Society.

Perhaps surprisingly, early observers did not immediately connect the presence of animalcules to food spoilage or disease. However, by the end of the seventeenth century, the essential tools were in place to understand the mechanisms of microbial food spoilage.

Louis Joblot, a professor of mathematics at the Royal Academy of Painting and Sculpture in Paris, described his work with microscopes in a book published in 1718 [10]. His instruments ranged from simple mounted lenses to more complex multi--lens designs. Using these tools, he observed bacteria, he referred to them as "fishes" or "insects," in hay infusions. In one experiment, he boiled the infusions for 15 min and then placed the hot liquid into two identical containers. He sealed one before it cooled and left the other open. Within a few days, the open container teemed with bacteria, while the sealed one remained clear.

In these simple experiments—conducted just a few years after van Leeuwenhoek's first observations and nearly a century before Appert's canning—Joblot had established the essential scientific principles behind food preservation: microorganisms exist in the environment, can be killed by heat, and do not spontaneously regenerate in sterile conditions unless exposed to air. Yet despite the significance of his findings, his work was not widely recognized.

In contrast, chemistry developed as a vibrant discipline over the course the eighteenth century. By the time Appert's methods were being evaluated, chemistry seemed the most natural framework for understanding them. Gay-Lussac had already conducted studies on fermentation, interpreting the process in chemical terms, so his chemically grounded approach to Appert's preserved food was consistent with prevailing scientific thought.

Chemists across Europe continued to develop Gay-Lussac's idea of a "ferment" produced by the reaction of oxygen with food components. However, as microbiologists began conducting their own experiments, their findings increasingly challenged this chemical interpretation. One good example is the work of the German physician, Theodor Schwann [6, 11]. Schwann sealed an aqueous nutrient solution in a glass globe and heated it in a Papin digester for 15 min—a process closely resembling modern canning. This was a more rigorous version of the experiments conducted by Joblot and Gay-Lussac, and once again no growth occurred. Schwann recognized, however, that the outcome could be interpreted in two ways: either the heat had killed any microorganisms present, or it had caused a chemical reaction with the air inside the globe, preventing it from forming the active ferment suggested by Gay-Lussac.

To test these possibilities, Schwann modified his experiment. He allowed air from the atmosphere to enter the heat-sterilized infusion, but only after it had passed through a red-hot glass tube. This treatment would not alter the chemical composition of the air, but it would destroy any microorganisms present. Once again, no spoilage occurred. This new finding could no longer be explained by the theory of a purely chemical ferment, lending support to the emerging alternative of microbial spoilage.

Schwann's experiment would seem to be compelling evidence that something living in the air was responsible for spoilage, and that thing could be destroyed by heat. Studies in the 1850s and 1860s went on to show that simply filtering the air through cotton wool could have the same effect as heating it; the "something living in the air" was a particle. Schwann went on to use the improved microscopes available in his time to see the yeast cells (which he called "sugar fungus"—later Latinized to *Saccharomyces*) present in fermenting grape juice and could even see bubbles of gas forming from them.

To us, the accumulating weight of evidence seems overwhelming, but, at the time, the idea that living organisms could be responsible for chemical changes was still considered absurd by many influential scientists. Among the skeptics were the German chemists Justus von Liebig and Friedrich Wöhler who mocked the idea that fermentation depends on living yeast cells in a satire: "it can be observed that the animals absorb the sugar from the solution from the very moment in which they escape from the eggs; the sugar can very clearly be seen to reach the stomach. It is digested instantaneously, and presently and with the greatest certainty, this digestion can be recognized as such, by virtue of the discharge of excrements. In a word, these infusoria feed on sugar, discharge spirits from the intestinal canal, and carbonic acid from the urinary apparatus. When full, the bladder assumes the form of a champagne bottle; empty, it is a little knob. After a little practice, one can observe that a gas bubble forms in its interior and grows tenfold" [12]. It would take the experiments, social connections, and forceful personality of Louis Pasteur to make the importance of microorganisms undeniable.

7.5 Pasteurism

Louis Pasteur made his scientific reputation as a chemist [11, 13–17]. He noticed the crystals of tartaric acid occurred in two distinct shapes, each the mirror images of the other. He then used a microscope and tweezers to painstakingly separate them into two piles and found the solutions he made from each affected light to the same degree but in opposite ways.[6] From this, he argued that although they were solutions

[6]Light can be polarized by passing it through a special filter, so that all the oscillations of its electromagnetic field are in one plane. Many sunglasses reduce the glare using this sort of polarizing filter. If you next pass the polarized light set at right angles to the first filter, all the light is blocked as the first filter caught the "up-and-down" components and the second filter caught the "side-to-side components." If you have two sunglass lenses, you can show this effect by looking through them both and seeing the brightness change as you rotate one of them. Next imagine introducing a sample between the two filters. If that sample rotated the plane of polarization of the light, turning the "side-to-side" components into something on a diagonal, then some of that light could make it through the second filter. If you wanted to measure the amount of rotation caused by the sample, you could simply rotate the second filter until it met the diagonal angle of the rotated light, and

of the same chemical, the molecules existed as mirror images of one another (i.e., chirality).[7] He noticed that the products of fermentation tended to be in one mirror form or the other, while similar molecules made in the lab were a mixture of both. This insight that the processes of life could leave a fingerprint on the molecules involved led Pasteur to shift his focus from chemistry to biology.

Pasteur began his work on fermentations in 1854 at the request of the father of one of his students, a factory owner who made industrial alcohol from sugar beets.[8] He used a microscope to show that the shape of the fermenting cells correlated with the quality of the fermentation. Plump, round budding cells (in modern terms, yeast) gave good quality alcohol, while smaller, rod-shaped cells (in modern terms, lactic acid bacteria) gave a sour taste. He also observed that transferring even a small amount of material from one fermentation to a fresh, unfermented solution could reproduce the characteristics of the original, for example, either a clean, quality fermentation or defects like ropiness and sourness. Although the microbiologists had not yet convinced the chemists, their capacity to solve practical problems was attracting the interest of industry, and Pasteur was a rapidly rising star in the French scientific community.

When Pasteur continued his investigations into the microbiology of fermentation, he approached the subject with a chemist's intuition and methods. According to the prevailing chemical view, if yeast acted simply as a catalyst, it should remain unchanged during the conversion of sugar to alcohol. But Pasteur's careful measurements told a different story. He observed that yeast gained mass by absorbing carbon from the sugar, and he also demonstrated that a nitrogen source was essential for fermentation to proceed. During the reaction, nitrogen disappeared from the solution and was taken up by the yeast. This made fermentation clearly part of yeast metabolism and not some sort of inert catalytic "ferment" the chemists still favored. He explicitly put his views in sharp contrast to those of the chemists: "I believe that alcoholic fermentation never occurs without either the simultaneous organization, development and multiplication of cells or the continued life of cells already formed. All the results in this paper seem to me completely in opposition to the opinions of Liebig and Berzelius" (quoted by Barnett [11]). But still not everyone agreed.

The last great defense of the spontaneous generation of life was led by the distinguished French naturalist Félix Archimède Pouchet [18, 19]. When he entered the fray in 1858, Pouchet, then 57, was significantly older than Pasteur, then 35, and he

---

once again nothing came through. Pasteur showed the angle of rotation of plane polarized light from tartrate solutions depended on concentration but was in opposite directions for the two mirror-image crystals.

[7] Pasteur demonstrated his achievement to the distinguished crystallographer Jean-Baptiste Biot who, as a young man, had accompanied Gay-Lussac on his balloon expedition. On seeing Pasteur's results, Biot is reported to have exclaimed: "My dear child, I have all my life so loved this science that I can hear my heart beat for joy" [17]. Scientific authority had passed to a new generation.

[8] Cultivation of sugar beets had been encouraged in France during the Napoleonic Wars in response to the British blockade cutting off access to Caribbean sugarcane plantations. Indeed, one of the advantages the government saw in Appert's invention was it allowed the preservation of food without the use of expensive sugar.

was unconvinced by the younger man's theories and methods. To Pouchet, what Pasteur called a theory was mere speculation, and his experiments were abstractions with little relevance to the real world. Critically, Pouchet did not fully grasp how rapidly microorganisms can reproduce; if only a single bacterium falls into a food and the conditions are right, there could be millions present the next day later. Because Pouchet misunderstood this, he found it implausible that air could contain enough germs to cause fermentation. He was left with the self-evident fact that living yeast cells must emerge from inert biological matter—the spontaneous generation of life.

When Pouchet attempted to replicate some of Schwann's experiments by introducing heated air into sterile media, he obtained the opposite results: his flasks began to ferment. To him, the only reasonable explanation remaining was spontaneous generation. However, when he presented his findings to the scientific elite of France at the Academy of Sciences in Paris, they were not convinced. While Pouchet's "common sense" interpretation resonated with many educated French citizens, much of the scientific establishment had already firmly embraced Pasteur's position that fermentation was caused by living yeast cells. It was possible Pouchet had made a mistake, after all the new sterile techniques necessary for microbiology were difficult to master. Perhaps the old naturalist simply lacked the necessary expertise and had somehow inadvertently contaminated his samples?[9] Still, the Academy took the matter seriously and offered a cash prize to anyone who could design and carry out experiments that would definitively resolve the issue.

By focusing on evidence from experimentation, the Academy was playing to Pasteur's strengths, and he took up the challenge enthusiastically. First, he followed Schwann and drew laboratory air through a cotton plug and then used microscopy to show how many microorganisms had been caught. He argued that these germs suspended in air could easily be a source of contamination if they came into contact with sterile media.

Next, he devised a now-famous experiment. He sterilized a flask of broth by boiling it, then softened the glass neck of the flask in a flame, and drew it out into a long, narrow, twisting "swan neck" that remained open to the air. This design allowed unheated air to reach the broth but prevented contamination because any airborne particles settled on the inner walls of the curved neck before reaching the liquid. These flasks did not ferment. As a control, he used identical flasks with their necks snapped off allowing immediate exposure of their contents to the air, and these did ferment. The contrast between the two outcomes provided compelling evidence against spontaneous generation.

Finally, Pasteur argued that if, contrary to his views, oxygen in air alone could catalyze fermentation, then any source of air would do. On the other hand, if microorganisms were the issue, then exposure to cleaner air should be less likely to cause

[9] Pasteur later showed that the mercury barrier that Pouchet had used to separate his sterile media from the air could itself be contaminated with suspended microorganisms. However, this was not known at the time, and the high standards of proof demanded of Pouchet reflected the Academy's prior skepticism of both the researcher and his claims.

fermentation. He boiled more flasks of media to sterilize them and then drew out their necks and sealed them, while they were still hot, creating a partial vacuum inside when the contents cooled—a close parallel to the process of canning. He then carried the sealed and sterile flasks high up into the Jura mountains, where he believed the air would be clean and contain fewer microorganisms. Once there, he broke open the neck with sterile pincers. Air, and whatever might be carried with it, was drawn into the flask, and then the neck was quickly resealed with a flame. His hypothesis was confirmed—the higher up the mountains he was when he opened the flasks, the fewer of them fermented. In November 1862, Pouchet withdrew from the competition, and Pasteur was awarded the 2500-franc prize. But the older man was not defeated.

The following year, Pouchet and his friends returned to the Academy with more evidence. They had replicated Pasteur's "mountain air" experiment and found the opposite results—two flasks opened at the bottom of the mountain fermented as expected, but the two opened at the top did as well. According to Pouchet's results, air was the important thing, not any living microorganisms in it.

Pasteur and the Academy were frustrated that Pouchet continued to challenge a "closed" issue and again criticized his statistics and his methods. *How many flasks did he see ferment? Only two—well, that's very few, had any others been opened that did not ferment? How had he opened the flasks? Using a file rather than pincers—well, that would be much harder to sterilize properly.* Replication was one thing, but it had to be done according to the exacting standards set by Pasteur and his friends at the Academy. Pasteur's earlier results that aligned with the Academy's expectations had faced far less scrutiny.

Pouchet vowed to return to the mountains with more flasks and with representatives of the Academy present to ensure good technique, but he never did, and Pasteur was left victorious. However, the issue might not have been so easily resolved if Pouchet had indeed gone back and repeated his experiment. Pasteur preferred to study yeast, which was easily killed by boiling. In contrast, Pouchet studied infusions of hay, possibly including organisms like *Bacillus subtilis*, which forms spores resistant to boiling. *Bacillus subtilis* grows best in the presence of oxygen, so his boiled flasks may have easily contained surviving spores that germinated and began to grow once the flasks were opened—even at high altitudes. This would have appeared to confirm his ideas about spontaneous generation.

Even in this alternate reality, it seems unlikely that Pasteur would have abandoned his theory—even if the boiled hay infusions had begun to ferment. He would likely have quickly identified the cause. Still, the scenario highlights that the standards of experimental proof required to settle a scientific question were not firmly established in advance. In the end, further experiments proved unnecessary. By 1872, the last major defender of spontaneous generation had died, and those left who still supported the idea were largely dismissed as cranks.

At the time of this famous debate, the real fight over spontaneous generation was already over. Pouchet was a scientist, but he was only a botanist from Rouen, and certainly not at the center of European science. By the time he got to the Academy of Sciences in Paris, he was the outsider, and the standards he had to meet were set by Pasteur and his admirers. If he had brought similar data a couple of decades earlier or made his claims somewhere else, where Pasteur was not so influential, his reception might have been different.

The sociologist of science Bruno Latour described what happened in Pasteur's laboratory as a "theater of the proof" [20]. According to Latour, "Nobody knew what spontaneous generation was; it had given rise to a highly confusing debate. But an elegant, open, swan-necked bottle, whose contents remained unalterable until the instant the neck was broken, was something spectacular and 'indisputable'." Pouchet's broad common sense view—living cells can form spontaneously from decaying organic matter—was constrained in Pasteur's experiment so all that mattered was what happened in these particular flasks [21]. By reducing the complex issue to a simple experiment, Pasteur was setting the rules of a competition he could definitively win. In Latour's view, the laboratory provided the setting where the drama could be acted out for an audience.

Pasteur continued to work on practical food microbiology throughout the 1860s and 1870s with seminal studies first on wine and then on beer. He studied wine spoilage while on holiday in eastern France in the summer of 1864 and there showed that a brief heating to 50 °C or 60 °C (122 °F or 140 °F) had a remarkable preservative effect on the local Jura wines.[10] Only later, when he came to write up his work, did he realize that Nicolas Appert had described a similar process in his book and was careful to give him credit for his prior art. However, Appert had had no firm theory of why his method worked, while Pasteur's discovery was valued practically by industrialists—it was also consistent with mainstream scientific understanding. Pasteur never studied on canned foods directly, although he clearly understood how they worked. The first scientists who did work on canning in a way that benefitted the canners were Americans—but that would be almost 30 years later.

7.6 A Paradigm Shift in the Science of Canning

Boyle and Papin outlined a scientific theory of food preservation based on preventing exposure to air, but their theory was incorrect. Although their theory was grounded in their experiments, their assumptions were rooted in ancient beliefs that "bad air" caused disease and decay. Without any concept of microbial life, they couldn't imagine an alternative explanation for what they saw. As a result, no matter how carefully they conducted their experiments, they were unable to properly test

[10] The process that bears Pasteur's name was only applied to milk by Soxhlet in the 1880s, although once again technological practice of milk preservation by heating occurred before scientific understanding [30]. Appert described a process for preserving condensed milk in his original 1810 *Book for All Households*—in the 1820s, public health campaigners recommended heating milk before serving to infants, and, in 1853, Gail Borden patented a popular process for canning sweetened condensed milk.

the "bad air" hypothesis. This view became the dominant scientific understanding and remained so until Pasteur ultimately discredited it by publicly defeating Pouchet.

The demise of the "bad air" explanation is an example of what the philosopher of science Thomas Kuhn identified as a paradigm shift in his classic *Structure of Scientific Revolutions* [22]. According to this view, at one time, most scientists shared a common paradigm: fermentation was a chemical reaction involving oxygen. Over time, this view was increasingly challenged by observations it couldn't explain, for example, Schwann's experiments. What followed was a period of confusion, during which multiple competing interpretations emerged. This phase lasted nearly 150 years from Joblot's observation that microorganisms were widespread yet could be destroyed by heat, to the Pasteur-Pouchet debates of the 1860s. Eventually, the debate was resolved. The scientific community coalesced around a new paradigm—microorganisms are responsible for fermentation—which was thereafter seen as obviously correct.

It's difficult to look back on a paradigm shift from the other side and truly grasp how it felt to scientists at the time. From our perspective, it can seem that the chemists who defended the old paradigm were being willfully disingenuous by ignoring evidence that now appears obvious. It did not seem like that to them. The things we see as important evidence from our paradigm might not seem important, or even seem to be evidence at all, from another. For example, the implications of Leeuwenhoek and Hooke's studies are obvious now, but this was not the case at the time. One reason could have been that early microscopes were very difficult to use, and most people who tried would see nothing but smears and colors. They would simply have to accept Leeuwenhoek and Hooke's word for what they had seen.[11] Even if they could make the observations, there may have been uncertainty over how the things they saw could be interpreted.

[11] Hillaire Belloc described these frustrations in his 1920 poem "The Microbe":
The Microbe is so very small
You cannot make him out at all,
But many sanguine people hope
To see him through a microscope.
His jointed tongue that lies beneath
A hundred curious rows of teeth;
His seven tufted tails with lots
Of lovely pink and purple spots,
On each of which a pattern stands,
Composed of forty separate bands;
His eyebrows of a tender green;
All these have never yet been seen--
But Scientists, who ought to know,
Assure us that they must be so ...
Oh! let us never, never doubt
What nobody is sure about!

About half a century before Hooke published *Micrographia*, Galileo Galilei had used a telescope to observe the stars and planets. His observations were pivotal to supporting the heliocentric model of the solar system which eventually displaced the older geocentric model. Kuhn used this story as an example of a paradigm shift in science, but Paul Feyerabend, another philosopher of science, stressed the ways Galileo had relied on nonscientific arguments to advance his case [23]. One of these can help us understand why the observations made by Leeuwenhoek and Hooke were not fully appreciated at the time.

Today, we readily accept the things we see through a lens are simply magnifications of reality, but that was not the case in Galileo's time. The lenses he used had optical defects which produced colored fringes around the objects you saw through them. We now understand these as illusions caused by the lenses themselves, not properties of the objects being viewed. If necessary, we can even prove this using other scientific theories describing how optics and vision work. Galileo, on the other hand, had to convince skeptics his new magnified observations were real while the colors were illusionary, even though the theories required hadn't yet been developed. While telescopes in his time were readily accepted as useful tools to magnify visible terrestrial objects, it wasn't fully justified to use them to identify completely new cosmological phenomena.

The first observers of microscopic images must have been faced with a similar conundrum (Indeed, Galileo built one of the first compound microscopes.). The instruments might reveal amazing new details of a flea, but people already knew fleas existed. On the other hand, could anyone be certain that the tiny specs moving at the limits of resolution are really a previously unknown form of life, or could they simply be another illusion generated by the instrument? Consequently, the attention of early microscopists was focused on larger objects, for example, insect anatomy or blood circulation, rather than the bacteria living at the limits of their instruments.

But by the middle of the nineteenth century, microbiologists had produced compelling evidence that microbial life was real and had important consequences for the real world. This new idea and the methods that supported it allowed them to design experiments that properly tested the bad air hypothesis, and they found it wanting. This was the period of most intense confusion, what Kuhn called a crisis in science, when there were communities of scientists who saw the problem in different ways and had data to support their views. However, each set of data came from different scientific paradigms each with their own methods, so evidence from one side did not serve to convince the other. When Pasteur finally won his decisive scientific victory, he did so by confining the contest to a domain where his rules applied.

Indeed, it is striking that the influential chemists defending the old paradigm felt confident enough to mock the microbiologists did so in very nonscientific terms and without feeling obliged to produce their own data to challenge the emerging ideas. They were frequently older and had made their reputations before the new methods came along. Perhaps they did not fully understand the new science, either because they chose not to engage seriously with ideas which they found ridiculous or perhaps because it was difficult to "unlearn" what they already knew. It would certainly have been a painful climbdown to admit that they, the experts, had been wrong. Science may be self-correcting but scientists, who after all are just people, are not.

Even now, when the conflict is apparently resolved, it is doubtless possible to find someone who still believes in the doctrine of spontaneous generation. The Internet is, after all, a big place. The new paradigm, once established, does not require unanimity, but any scientist who publicly holds the old position would be seen as either troublingly eccentric or, more likely, not really a scientist after all.

7.7 Reflections for Modern Scientists: Getting Better at Being Wrong

The American Society for Microbiology's introductory textbook on food microbiology describes the end of the spontaneous generation debate as follows: "Pasteur disproved this using a simple but elegant experiment with curved-neck flasks … By showing that 'life comes from life' he killed the idea of spontaneous generation. From [these results], he postulated that fermentation was a biological rather than chemical process" [24].

When we train to be scientists, we learn some version of the scientific method which includes an understanding of which theories and techniques are appropriate in our field and what sort of evidence is needed to support a conclusion. While the details of what counts as "scientific" and what counts as "method" might vary, a scientist is expected to be expert in what holds in their particular field. Scientists are also expected to share the belief that progress is a result of the correct application of these methods, we had a false belief and then, after doing the right experiment, we now know the truth. The story in the microbiology textbook is just one example of senior scientists trying to instill this worldview in trainees. However, as we have seen in this chapter, the story isn't true.

Pasteur's swan-necked flask experiments simply marked the end of a decades-long debate; he wasn't postulating anything new; and most participants were already convinced of the answer. Indeed, it's hard to point to anyone, not even poor Pouchet, who actually changed their minds based on the results. Science is done by people influenced by the same morass of social factors and historical circumstances as everyone else.

So, if our notion of rational scientific progress is so badly flawed, should working scientists keep acting as if it were true and keep stressing its importance to trainees? With some qualifications, I think they should.

One reason to continue to value the ideal of rational progress in science is that while paradigm shifts are important, they are rare. Over the course of their careers, very few scientists are likely to witness anything as transformative as the nineteenth-century move from chemical to microbial explanations of fermentation. Most will work in what Kuhn called a period of "normal science" where what counts as valid methods and theories are already established [22]. The philosopher of science Laudan has argued that, under these circumstances, the sensible thing is to follow the current paradigm, not because we are certain it is true but because that's the one shown to be most effective in solving problems [25]. The few scientists who lead a successful paradigm shift and establish a new era of productive normal science are remembered as heroes or even as founders of disciplines; but most who try fail and are relegated to the fringes as cranks.

In any case, scientific ideas can still be useful even if they are wrong. Gay-Lussac was wrong when he argued that foods were preserved by a vacuum, but his theory was believed by many canners well into the twentieth century—long after the scientific controversy was resolved [26]. One reason for this is the vacuum had already proved useful for other reasons.

The London canners (Chap. 3) protected the seams of their new tin cans by venting them during cooking, allowing some of the air and steam to escape and reducing the internal pressure. They then sealed the vent hole with a drop of solder before finishing the cook. As the sealed cans cooled, their contents contracted, creating a partial vacuum inside. They quickly observed that cans with concave ends were properly sealed but ones with flat ends would spoil. It wasn't the absence of a vacuum that caused the spoilage, but rather the microorganisms that were drawn in along with the air through tiny defects in the seams as the can cooled. In fact, modern canneries still create an internal vacuum either by sealing cans under reduced pressure or by first flooding them with steam which condenses and contracts as the can cools after cooking. The vacuum remains an essential element in canning, even though Gay-Lussac's scientific explanation for it was incorrect.

A second reason for scientists to continue valuing rational scientific progress is to see it as an ideal they are willing to strive for, rather than a reflection of current practice. Like any human endeavor, science has flaws ranging from publication bias and failures in replication to outright fraud. Sociologists might reasonably argue that these issues further illustrate how scientific "truth" is shaped by a network of social influences (this will be a focus of Chap. 10). Yet for scientists who treat rational progress as an aspirational goal, such shortcomings serve as motivation to improve the quality, transparency, and integrity of their ongoing work. The ways we learn from historic examples can support each of these perspectives.

Consider three different reactions we, as scientists working today, might have to the stories of the German chemists resisting the new ideas from microbiology in the early nineteenth century. First, we could simply criticize them for being so obviously wrong. Certainly, the temptation toward this interpretation is strong; how could they see this evidence and still not reach the obvious conclusion? However, this isn't a helpful lesson. They were leaders in their field whose achievements are likely to be remembered far longer than anything we will do. While we can now see their mistake in hindsight, we are unlikely to do much better at recognizing our own errors today.

Alternatively, we could simply see this story as yet another example that what we accept as scientific truth is simply a social construct without a firm rational basis.

Again, this interpretation has merits, but again it isn't helpful. If science is really so flawed, there can be no point in going to the lab tomorrow.

A third and more productive reaction for working scientists is to look at the story of the German chemists as a cautionary tale. If our focus is not on their error, or even that error is inevitable, but rather *how* they went wrong, then we can find lessons from their mistakes and try to avoid them ourselves.

The German chemists believed their views were based on experimental results and sound scientific reasoning, yet they ignored relevant observations from other scientists. Worse, they expressed their disagreement as satire rather than as rational scientific argument. We probably aren't smarter than they were, but we can learn to recognize these types of behavior as misguided and try to avoid them. Consider all relevant evidence—even if it was done by someone else. When you do disagree with another scientist, do so in terms of rational argument. Do whatever you can to bring your own contributions as a scientist closer to the rational ideal.

References

1. Thomas, O. (trans.): *Geoponika*, vol. 2 (1806)
2. Tulchinsky, T.H., Varavikova, E.A.: A history of public health. In: *The New Public Health*, pp. 1–42. Academic Press (2014). https://doi.org/10.1016/B978-0-12-415766-8.12001-4
3. Cowell, N.D.: An Investigation of Early Methods of Food Preservation by Heat. University of Reading (1994)
4. Gay-Lussac, J.L.: Extrait d'une memoire sur la fermentation. Ann. Chim. **76**, 245–259 (1810)
5. Gillispie, C.C., Holmes, F.L.: Dictionary of Scientific Biography. Scribner (1981)
6. Barnett, J.A.: A history of research on yeasts 1: work by chemists and biologists 1789–1850. Yeast. **14**, 1439–1451 (1998)
7. Boghurst, W.: An Account of the Great Plague of London in the Year 1665. Shaw and Sons (1894)
8. Wollman, A.J.M., Nudd, R., Hedlund, E.G., Leake, M.C.: From Animaculum to single molecules: 300 years of the light microscope. Open Biol. **5**, 1–10 (2015)
9. Lane, N.: The unseen world: reflections on Leeuwenhoek (1677) 'concerning little animals'. Philos. Trans. Biol. Sci. **370**, 1–10 (2015)
10. Lechevalier, H.: Louis Joblot and his microscopes. Bacteriol. Rev. **40**, 241–258 (1976)
11. Barnett, J.A.: A history of research on yeasts 2: Louis Pasteur and his contemporaries, 1850–1880. Yeast. **16**, 755–771 (2000)
12. De Mayo, P., Stoessl, A., Usselman, M.C.: The Liebig/Wöhler satire on fermentation. J. Chem. Educ. **67**, 552–553 (1990)
13. Schwartz, M.: The life and works of Louis Pasteur. J. Appl. Microbiol. **91**, 597–601 (2001)
14. Martensen, R.L.: Louis Pasteur (1822–1895): his laboratory and the world. JAMA. **274**, 995 (1995)
15. Ligon, B.L.: Louis Pasteur: a controversial figure in a debate on scientific ethics. Semin. Pediatr. Infect. Dis. **13**, 134–141 (2002)
16. Vallery-Radot, R.: The Life of Pasteur. Doubleday (1912)
17. Holmes, S.J.: Louis Pasteur. Harcourt, Brace and Company (1924)
18. Farley, J., Geison, G.L.: Science, politics and spontaneous generation in nineteenth century France: the Pasteur-Pouchet debate. Bull. Hist. Med. **48**, 161–198 (1974)
19. Roll-Hansen, N.: Experimental method and spontaneous generation: the controversy between Pasteur and Pouchet, 1859–64. J. Hist. Med. Allied Sci. **34**, 273–292 (1979)

20. Latour, B.: The Pasteurization of France. Harvard University Press (1988)
21. Schaffer, S.: The eighteenth Brumaire of Bruno Latour. Stud. Hist. Philos. Sci. Part A. **22**, 175–192 (1991)
22. Kuhn, T.: The Structure of Scientific Revolutions. University of Chicago Press (1996)
23. Feyerabend, P.: Against Method. Verso (1993)
24. Matthews, K.R., Kneil, K.E., Montville, T.J.: Physical methods of food preservation. In: *Food Microbiology: An introduction*, ASM Press, Washington DC, pp. 489–506 (2017)
25. Laudan, L.: Pogress and Its Problems: Towards a Theory of Scientific Growth. University of California Press (1977)
26. Bitting, A.W.: *The Canning of Foods; A Description of the Methods Followed in Commercial Canning. USDA Bureau of Chemistry – Bulletin No. 151*. USDA (1912). https://doi.org/10.5962/bhl.title.30726
27. Cowell, N.: Who invented the tin can? – a new candidate. Food Technol. **49**, 61–64 (1995)
28. Cowell, N.D.: More light on the dawn of canning. Food Technol. **61**, 40–45 (2007)
29. Bulloch, W.: The History of Bacteriology. Oxford University Press (1936)
30. Holsinger, V.H., Rajkowski, K.T., Stabel, J.R.: Milk pasteurisation and safety: a brief history and update. Rev. Sci. Tech. **16**, 441–451 (1997)

Chapter 8
The Birth of a Science of Canning

Abstract Wisconsin and Massachusetts 1894–1902. A growing canning industry meets a new generation of academic scientists trained in microbiology and motivated to help. The scientists quickly translate their experience from other contexts and provide useful results which are disseminated by trade across the industry. The social and political frameworks needed for collaboration.

8.1 The Science of Canning in Wisconsin

The Landreth family had been selling seeds in Bucks County, Pennsylvania, since the time of the Continental Congress. In 1875, they sent 17 year old Albert Landreth west to Wisconsin to manage a new investment: 200 acres of peas in Manitowoc County along the shores of Lake Michigan [1, 2]. In 1883, he set up an experimental processing operation in a back room of his mother-in-law's hotel. By 1887, it had grown into the first full-sized cannery in the state, and Landreth hired Francis Patterson to supervise the crucial cooking stage. Patterson was a "processor," much like Andrew Hapgood had been in the Sacramento salmon cannery (see Chap. 5). Processors were highly valued at the time for their expert knowledge, which they guarded fiercely—sometimes even locking the door to prevent the cannery owner from learning the precise times and temperatures required for each product.

However, Patterson's vaunted expertise was lacking, and, by 1894, the Landreth cannery was facing a problem that had plagued canners since the very beginning. A few days after cooking, some of the cans would first swell and then explode. If their expert couldn't solve their problem, who else could they turn to for help?

Harry Luman Russell was born in the tiny farming village of Poynette, Wisconsin, the son of a local doctor.[1] In 1884, he made the short journey to Madison to study at the University of Wisconsin. There, he was inspired by a group of young faculty

[1] Russell's life and early achievements in canning are described most vividly by May in his 1937 history of the industry [1]. Hastings [3] and Fred [4] offer a more thorough discussion of his life and scientific work, and Beardsley [5] describes his influence on the University of Wisconsin.

J. Coupland, *Containing Nature*, https://doi.org/10.1007/978-3-032-11882-0_8

members who were beginning to introduce the emerging science of bacteriology into the curriculum. Motivated by their influence, Russell continued his studies in the field and remained at the university to pursue a master's degree in bacteriology.[2]

At the time, most American universities focused primarily on undergraduate teaching, so many ambitious young science graduates sought part of their research training in Europe—especially in Germany. Harry Luman Russell followed this path. Between 1890 and 1892, he studied under Robert Koch in Berlin and at the Pasteur Institute in Paris. He then returned to the United States to complete his PhD at Johns Hopkins University in Baltimore, which was considered the most modern research university in the country at the time. After earning his doctorate, Dr. Russell returned to his alma mater as an assistant professor of bacteriology.

The University of Wisconsin was founded in 1848—the same year Wisconsin joined the Union—and received its federal land grant under the Morrill Act in 1862 [7–9]. The land grant provided funding but required the institution adopt a clearly defined practical mission: "to teach such branches of learning as are related to agriculture and the mechanic arts, in such manner as the legislatures of the States may respectively prescribe, in order to promote the liberal and practical education of the industrial classes in the several pursuits and professions in life." The university expanded its role after the passage of the federal Hatch Act in 1887 provided $15,000 to establish an agricultural experiment station with a mission to conduct research and share results in ways that would benefit the state's farm economy. It was through this experiment station that Russell was hired, in the hope that he could help reduce the impact of cattle diseases, especially tuberculosis which was a significant concern in Wisconsin. However, when the local pea canners showed up looking for help, their problem landed on Russell's desk [10].

Russell applied the microbiological techniques he had learned in Europe and at Johns Hopkins to the issues at the Manitowoc County cannery. He took samples from cans and inoculated them into sterile nutrient media. If anything grew, it would indicate the presence of live bacteria. The intact cans showed no growth, but the exploded ones were heavily contaminated with bacteria which were clearly visible under the microscope. More tellingly, these bacteria could reproduce without oxygen and produced gas as they multiplied. This made them prime suspects: if they behaved the same way inside sealed cans, the gas could cause the cans to swell and eventually burst.

Still, all Russell had at this stage was a correlation—swollen cans contained the bacteria, while intact cans did not. It was possible that some unknown factor both caused the swelling and encouraged bacterial growth, without the bacteria being the

Russell's papers on canning are reproduced by Bitting in *Appertizing* [6]. His influence on the pea cannery is described by Elhert [2]. These sources form the basis for this section.

[2] In his thesis, Russell showed the ice from Lake Mendota on the university campus did not contain pathogenic bacteria, an important practical concern for the commercial ice-harvesting industry.

root cause. To prove causation, Russell needed to go beyond observation and design an experiment.

Fortunately, a similar question had recently been addressed in the field of medical microbiology [11]. Scientists could isolate many types of bacteria from sick individuals—but how could they determine which of them was responsible for the illness? Russell's German research mentor, Robert Koch, had already proposed a framework to answer this question. To establish causation, three conditions had to be met:

(i) The microorganism must be found in sick individuals but not healthy ones.
(ii) The microorganism must be isolated and grown in pure culture and then introduced into a healthy host, which should develop the disease.
(iii) The same microorganism must then be re-isolated from the newly infected host.

This framework became known as Koch's postulates. While some diseases—such as most cancers, those caused by viruses, or those with asymptomatic carriers—do not fit these criteria, many others do [12].

Russell applied his mentor's methods for human and animal disease to canned foods—what was making the cans "sick"? He selected "healthy" intact cans, those that hadn't shown the symptom of swelling, and carefully sterilized their outer surfaces with a flame. Then, using a sterilized drill bit, he bored a hole into the metal, introduced a small amount of the gas-producing bacteria isolated from the "sick" swollen cans, and resealed the opening with a drop of solder. Within 36 h, the inoculated cans started to swell. When opened, they were found to be heavily contaminated with the same bacteria. This was no longer just a statistical association—it was a clear demonstration of cause and effect, established through controlled experimentation.

Not only did he identify the source of the problem in terms his scientific peers could understand—describing the culprit as "an organism capable of fermenting sugar solutions with a copious evolution of gas" and "able to live either in the presence of or without free oxygen" [10]—he also offered a practical solution that canners could use. One option was to simply extend the cooking time to kill the bacteria, but this tended to make the liquid in the can cloudy and unattractive. Instead, Russell showed that increasing the temperature from 111 to 117 °C (232 °F to 243 °F) by increasing the pressure in the retort while keeping the cooking time the same (30 min), could prevent spoilage without compromising product quality.

While the processors like Francis Patterson had jealously guarded their canning expertise, Russell took a different approach and published his findings as a report from the Wisconsin Experiment Station [10]. These reports were shared among the other state experiment stations, although it is not clear how widely they were read by canners. Historians Gabriella Petrick [13] and Anna Ziede [14] suggest Russell had limited influence beyond those canners he worked with directly. They, however, remembered "Dean Russell" as a Wisconsin hero [1]. Despite this local recognition, canning research in Wisconsin did not develop further. The state only had three canneries, and Russell's real interests lay in cattle diseases. A sustained academic research program in the science of canning would not begin until the following year, in Boston.

8.2 The Science of Canning in Massachusetts

Samuel Cate Prescott was born the son of a blacksmith in South Hampton, New Hampshire, in 1872.[3] In 1890, he enrolled at the Massachusetts Institute of Technology (MIT) in Boston, where he pursued studies in science, with a particular focus on bacteriology and chemistry. His bacteriology instructor was William Sedgwick—a microbiologist interested in public health work and a founder of the American Society for Microbiology—and his chemistry instructor was Ellen Richards, the first female faculty member at MIT and a pioneer in both women's scientific education and community nutrition [19]. Both of them were important mentors to the young scientist, and, as we shall see later, both were emblematic of the type of university he was attending.

Prescott graduated from MIT in 1894 with a degree in chemistry.[4] He worked briefly on wastewater treatment in Worcester, Massachusetts, but almost immediately returned to the university as Professor Sedgwick's assistant. Late the following year, Sedgwick was approached by William Lyman Underwood of the Underwood Company of Boston about an intermittent problem they had been having with their canned clams.

The Underwood Company was founded by William Lyman Underwood's grandfather, also named William. The older William Underwood had learned food preservation as an apprentice in London during the earliest days of British canning before emigrating to America in 1817 [15, 20]. In the 1820s, he established one of the first canneries in the United States processing tomatoes, corn, and seafood for export and to supply westward-bound migrants.[5] The company prospered through military contracts during the Civil War and, in that period, developed its signature devilled ham—now the only surviving product carrying its brand into the twenty-first century. William's son, yet another William, took over the family business on his father's death in 1864, and eventually his three sons, William Lyman along with his brothers Henry Oliver and Loring, joined the family firm.

[3]Details of Prescott's life are taken from Underwood's recollections of their work together [15] and from a biography written by his student, Samuel Goldblith [16]. Goldblith is an interesting character in his own right and deserving of a detailed biography not just an extended footnote. After graduating from MIT in food technology, he served in the Second World War and was captured by the Japanese [17]. While in a prison camp, he conducted experiments to track the deterioration of the eyesight of his fellow prisoners due to the malnutrition they were all suffering [18]. After the war, he returned to MIT as a faculty member and department chair in food science. A student who worked with him describes him as "very kind and supportive." Later in life he wrote some of the histories of canning cited here.

[4]Prescott's senior thesis was entitled *Salts as Nutrients for Bacteria*, suggesting bacteriology as well as chemistry remained important to him. Despite his future eminence as a researcher, he never earned a doctorate.

[5]The Underwood Company started their business packing food in glass bottles but soon moved on to soldered metal cans. They are believed to be the first people to have abbreviated the name for the metal canisters they made to "cans" [15].

The problem the Underwoods faced with their canned clams was similar to the one that had faced Landreth in Wisconsin. A few days after processing, some cans would swell and eventually explode, spraying blackened fragments of liquefied fish and releasing hydrogen sulfide—the unmistakable foul odor of rotten eggs. Sedgwick assigned the problem to his new assistant, Samuel Prescott.

Surprisingly, William Lyman Underwood didn't just delegate the issue or write a check to the university, but instead he became involved personally. He worked alongside Prescott in the lab most afternoons and sometimes even moved their laboratory into the factory. Prescott later recalled spending a summer with Underwood at a corn cannery in Maine, where they camped out and used incubators heated by candles to accelerate spoilage in test cans. Occasionally, they were jolted awake in the night by the dull thud of exploding cans, only to be scolded by the cannery manager in the morning for the mess their work had created. Through their collaboration, Underwood and Prescott developed a lifelong friendship, bonded by their shared curiosity, long hours, and mutual love of camping, fishing, and scientific discovery.[6]

Prescott and Underwood approached the canned clam problem much like Russell had done in Wisconsin with the peas. The swollen cans were teeming with bacteria which, when introduced into intact cans, caused them to swell and eventually explode after a brief incubation period. Prescott and Underwood identified the spoilage organisms—often around ten distinct types per food—based on their microscopic appearance and their growth patterns on various nutrient media in petri dishes. Underwood was a skilled photographer and illustrated their findings with striking images of bacterial colonies growing on plates, as well as individual cells viewed under the microscope.

Their experiments also showed that once previously sound cans had been inoculated with the bacteria from spoiled cans, even a full day of boiling in a water bath couldn't prevent them from spoiling. In contrast, just 30 min in a pressure retort was sufficient to preserve the cans indefinitely. Retorts had been in use in canneries since the 1870s and were known to be effective (Chap. 4). However, this was the first time science had explained why.

Prescott and Underwood quickly recognized that the important temperature for safe canning was not the temperature of the retort but the temperature at the center of the can. It was here, at the coldest point, where the last heat-resistant bacteria could potentially survive processing. If even one organism remained viable at the end of processing, it would be at that point in the can, and from there, it could

---

[6]Aside from his personal friendship with Prescott, and his evident love for the work, relationships within his family conspired to force Underwood from industry to academia [20]. His eldest brother, Henry, was a dominant personality who saw himself as the one who would lead the family firm into its third generation. By the end of the century, he had forced his younger siblings out. Lyman and Loren were wealthy and so, without the usual financial responsibilities, were free to pursue their own interests. Loring became a landscape architect, and William Lyman pursued his personal interests in wildlife photography and as a public lecturer and author as well as continuing with research and teaching on foods at MIT for the rest of his life. He died in 1929, and his company endowed a professorship in his name. The first recipient of this recognition was Samuel Goldblith.

multiply and spoil the food. Prescott and Underwood placed small registering thermometers inside cans to measure the maximum internal temperature reached during processing. Their experiments showed that cans that were larger or those that contained more solid pieces of food took longer to reach the target temperature. Based on their findings, they recommended a center temperature of 121 °C maintained for 10 min as a safe minimum standard for sterilization.

Prescott and Underwood published their findings, and the way they did so reveals much about their intended audiences.[7] They began by giving public lectures in Boston at meetings organized by the Society of Arts—an organization founded alongside MIT to share the university's work with the broader community. These talks were soon afterward published in MIT's house journal, *Technology Quarterly*, and, in one case, essentially the same material appeared in the prestigious academic journal, *Science* [21]. These publications helped Prescott establish his reputation both at his university and in the larger scientific community. At the same time, he and Underwood made deliberate efforts to reach the wider canning industry, ensuring their research had both practical and academic impact.

Within a few weeks of their first talk at the Society of Arts, William Lyman Underwood wrote to the trade journal *Canner and Dried Food Packer* to describe their work on clams and lobsters. In his letter, he asserted priority over a similar study recently reported by the Canadian government and added: "You may also be interested to learn that for nearly a year we have been investigating the bacteria conceived in the spoiling of corn, 'sour corn,' and hope to publish an article on this subject in a short time." This was a start of an ongoing relationship between Prescott, Underwood, and the canning industry—conducted largely through their trade associations. The scientists spoke at the canners' winter meetings in Buffalo in 1898, in Detroit in 1899, in Rochester in 1901, and in Milwaukee in 1902 (Prescott missed the 1900 meeting because he was visiting laboratories in Europe, and he instead sent a letter which was read into the minutes). The academics were interested in the canners, and the canners were interested in the academics.

In many cases, Prescott and Underwood's reports to the canners were simply reprints of their scientific papers from *Technology Quarterly*. However, in other instances, they added much more detailed discussions of cannery operations and offered practical guidance on how packers could benefit from the latest research. In 1902, their collected papers were published as a book, *Science and Experiment as Applied to Canning* [22].

The question-and-answer sessions following their talks were often transcribed and give a flavor of how the canners were responding to the latest results from the university. For example, this exchange from 1898 reveals an impressive level of practical processing knowledge from the academics and scientific expertise from the canners:

> Question: Are you satisfied that fifty-five minutes in a retort at 250 degrees [121 °C], that corn can be cooled down at once?

[7] All of these early papers are reproduced in Bitting's *Appertizing* [6].

Mr. Underwood: I think that would be dangerous because it takes fifty-five minutes to get that heat there. The quicker corn is cooled after it is thoroughly processed the better, Fifty-five minutes at 250 degrees will thoroughly cook it. I think that is the danger line. You must understand we are not corn packers.
Question: You contend that all corn contains bacteria?
Mr. Underwood: Yes.
Question: Is it due to the acid in tomatoes that it is easier to save tomatoes than it is corn?
Mr. Underwood: That is something we don't know anything about. Tomato is a liquid and the heat gets into the center of the can in a short time.

This growing recognition that scientific research could yield practical results led to the rapid development of other laboratories dedicated to solving problems in the canning industry [13, 14, 23]. Edward Duckwall, the son of a prominent Midwestern canner but with little formal scientific training, published *Bacteriology as Applied to Canning and Preserving* in 1899 [11]. In 1903, he opened an industry-funded canning research laboratory near the H. J. Heinz factory in Aspinwall, Pennsylvania. In 1907, the National Canners Association (NCA) was formed from the amalgamation of regional trade associations and opened a more substantial research laboratory of their own in Washington DC. This would later provide the scientific infrastructure for future industry-led efforts at solving the problems of botulism (Chap. 9) and tin in foods (Chap. 10).

Universities also developed applied research programs focused on the problems of the food industry. Samuel Prescott went on to a successful career as a researcher, teacher, and administrator at MIT [16]. After his mentor's, Professor Sedgwick, death in 1921, he took over as head of the Department of Biology and Public Health and eventually rose to be the first Dean of MIT's School of Science. In 1927, Prescott hired one of his former students, Bernhard Proctor, and supported him in launching the first courses, degree, and academic department of food technology at an American university. Prescott and Proctor coauthored one of the first comprehensive textbooks in food science in 1937 [24], and in 1939, they together helped found the professional organization for food scientists—the Institute of Food Technologists.

8.3 Why Scientists Started to Help Canners

When the prominent and politically connected scientist Joseph Louis Gay-Lussac was asked to evaluate Appert's invention, he applied his understanding of chemistry to offer a plausible explanation: canning worked by preventing oxygen from reacting with food and causing spoilage (Chap. 7). He was wrong—but that hardly mattered. Gay-Lussac's social prestige and scientific authority served as a kind of "blessing" for Appert who, in any case, didn't need a scientific theory to run his business. After that, there was no meaningful applied research on the science of canning until the stories described here.

In both Wisconsin and Massachusetts in the 1890s, it was the canners who approached the scientists with a specific and pressing problem. What followed was notable; not only did the scientists respond, but their response was collaborative rather than a one-way transfer of knowledge from the lab to the factory. Russell had learned his science from Koch, but he still needed to understand the canners' context and integrate that with his scientific training to find a solution. Prescott had to teach Underwood microbiology, but, in return, Underwood taught him the practical realities of industry. Both Russell and Prescott brought their laboratory equipment into the canneries to do experiments under real industrial conditions. This hands-on, reciprocal approach was key to their success.

The success of their research was measured not only by its reception in scientific circles but also by how effectively it was applied in industrial practice. Both scientists made a conscious effort to produce findings that were valuable to both scientific and industry audiences. Their reports included detailed bacteriological information for academic readers as well as practical recommendations for canners. The fact Prescott earned promotion suggests the university found his contributions significant. The fact that canners continued to invite Prescott and Underwood to their annual meetings suggests they found the work to be genuinely useful. The nature of the discussions during these sessions also reflected a mutual respect between scientists and industry professionals.

There are three reasons their achievements were possible in the 1890s but not in the 1810s: the science had developed, the missions of universities had changed, and the canning industry had grown. Each factor contributed to altering the social dynamics between scientists and industry and together allowed effective collaboration.

First, the science had developed. The work of Pasteur and others in the mid-nineteenth century led to a paradigm shift, and, by the 1890s, American scientists working on canning had been educated within this new framework. Prescott and Underwood opened their very first papers by outlining the basic bacteriological theory of canning as they understood it at the time. The story they told is essentially what would be found in a modern textbook. Before even starting work, Prescott and Underwood already understood that most bacteria in a can could be easily destroyed by heat, while others were more resistant. They also understood that the vacuum inside a can was useful for detecting leaks but not essential to food preservation. What is more, when Prescott and Underwood began their experiments, the methods they used were almost identical to those used by Russell the previous year in Wisconsin—even though they were unaware of his work at the time [25]. The fact that two independent groups approached the problem in the same way suggests this was simply the "obvious" method for research bacteriologists at the time. The science was fully in place before they even walked into a cannery, and all the academics had to do was make it work in a new context.

The second reason for the success of the 1890s research was because the academic institutions had also changed. Although Russell and Prescott worked at very different types of universities, both shared a focus on serving the needs of their surrounding communities. We have already seen how deliberate government policy, enacted through the Morell and Hatch Acts, had made the University of Wisconsin into an institution responsive to the practical needs of local industry. MIT, by contrast, was a private university founded in 1861 with a different mission [26]. Since the 1840s, its first president, William Barton Rogers, had been advocating for a technical university in Boston that would support the growing industrial economy, rather than simply replicating the traditional liberal arts education offered by existing institutions. Although Rogers was focused on industrial manufacturing, MIT was intended for young, middle-class men who would become factory managers—not for factory workers or skilled artisans. His practical yet class-conscious vision aligned well with the Progressive Era, a time when many believed that science and progress could solve society's endemic problems.

The careers of Prescott's key mentors—William Sedgewick and Ellen Richards—serve as examples of the ethos of MIT at that time. Sedgewick studied the growth of microorganisms in water and was engaged in campaigning for a water treatment system for the public supply in Lawrence, Massachusetts—the first of its type in the country. Richards was a prominent advocate of the "euthenics" movement, which promoted the rational improvement of living conditions as a better path to social progress than the selective human breeding being advocated by contemporary eugenicists [27]. She ran kitchens in the poorest parts of the city where she taught immigrant women how to feed their families healthy food on a very limited budget.

While government policy shaped the University of Wisconsin, MIT was shaped simply by the choices of its leadership and by the economic and social conditions of the time. Despite their very different institutional characters and histories, faculty members in both universities saw it as their responsibility to take their expertise out of the laboratory to try to solve problems in their communities.

The third and final reason that a productive relationship between scientists and industry was possible at the end of the nineteenth century was that the industry had changed. When Gay-Lussac visited Appert's factory in 1810, he was looking at the entire global extent of the canning industry. There was potential, but little immediate economic impact. By the time canners approached academics again in the 1890s, they represented parts of a large industry with real economic power and a growing sense of their shared interests.

Since the 1880s, equipment manufacturers had been organizing professional societies, journals, and meetings for canners. These were the meetings where Prescott and Underwood presented their work. The 1902 book where their papers were collected, *Science and Experiment as Applied to Canning*, was published by the Sprague Canning Machinery Company of Chicago and illustrated with pictures of the equipment they sold [22]. This growing sense of a collective "industry interest" would, by the 1920s, provide the momentum to build on Prescott and Underwood's early findings and address the problem of botulism in canned food (Chap. 9).

8.4 Reflections for Modern Scientists: Working with Industry

Vannevar Bush, writing at the end of the Second World War, argued fundamental science was the engine of technological development and economic progress. "New products and new processes do not appear full-grown," claimed Bush: "They are founded on new principles and new conceptions, which in turn are painstakingly developed by research in the purest realms of science" [28]. This was certainly not the case in the early decades of the canning industry; Appert was a chef and Donkin and Girard were engineers. While the scientific establishment might have helped them make productive connections, there was no useful science of canning (Chaps. 3 and 7). In this chapter, we saw at last beginnings of a productive relationship and a science of canning, but it was still not Bush's model of scientists passing their discoveries down to technologists who would use them to make new products and processes. Instead, it was collaborative.

Many universities today argue that they deserve public support because their researcher helps industry and, eventually, creates jobs and economic growth. There are many things that might limit the success of such an endeavor, but one factor that is at least partly under the control of an individual scientist is how they collaborate. To achieve this goal more effectively, they would do better to follow Russell and Prescott's example rather than Bush's vision. Russell and Prescott were well trained in the latest scientific methods, but that alone wasn't enough. What made the difference was their respect for the canners and their recognition that the canners' knowledge mattered. Building effective collaborations depends on building real relationships. One reason Prescott ultimately had a greater impact than Russell was his sustained social and professional engagement with the canning community. He spent time in their factories and regularly attended their meetings. Over many years, he came to understand their needs—just as they came to understand his capabilities. Through this ongoing exchange, they built trust and came to value one another's contributions.

However, many modern research universities do not adequately value the time-consuming work of building relationships. Success is often measured by high-impact publications and large grants, which in turn prioritizes relationships between scientists and major government funding agencies. When industry involvement does occur, it often comes late in the process—typically in the form of letters of support for grant proposals whose broad objectives have already been defined. Even the phrase commonly used for this sort of work at universities, "applied science," carries a sense of taking our science and applying it like a coat of paint over a passive surface. Instead, the ideal should always be collaborative, with knowledge moving backward and forward as academic and industrial partners co-create useful solutions. If university leaders genuinely see a role of their institution as supporting industry, they should look for ways to value the type of work and the time it takes.

We should also recognize that building collaborative relationships was challenging for the canners as well. From an industry perspective, sharing privately held knowledge is risky. Scientific research, on the other hand, even if it isn't done in the context of a public university, depends on the exchange of information in order to progress. Any effective collaboration that involves the cocreation of knowledge must find ways to manage this conflict, so both partners benefit enough to want to participate. Trade associations can play a crucial role in supporting this kind of partnership, as their goal is to benefit the industry as a whole, not just individual companies. Their involvement is essential in next stage of the developing science of canned food—the concerted effort in the 1920s to solve the issue of botulism.

References

1. May, E.C.: The Canning Clan. Macmillan Publishers Limited (1937)
2. Elhert, E.: Food Processing – A Manitowoc Triumph. Manitowoc County Historical Society (1970)
3. Hastings, M.: A review of the scientific work of Harry Luman Russell together with a list of his books and scientific papers. In: Papers on Bacteriology and Allied Subjects, pp. 9–29. Madison (1921)
4. Fred, E.B.: Harry Luman Russell, 1866–1954. J. Bacteriol. **68**, 133–134 (1954)
5. Beardsley, E.: Harry L. Russell and Agricultural Science in Wisconsin. University of Wisconsin Press (1969)
6. Bitting, A.W.: Appertizing. The Trade Pressroom (1937)
7. Curti, M., Carstensen, V.: The University of Wisconsin: A History. University of Wisconsin Press (1949)
8. Johnson, E.L.: Misconceptions about the early land-grant colleges. J. High. Educ. **52**, 333 (1981)
9. Key, S.: Economics or education: the establishment of American land-grant universities. J. High. Educ. **67**, 196–220 (1996)
10. Russell, H.: Gaseous fermentation in the canning industry. Wisconsin Agric. Exp. Stn. (reproduced by Bitting, 1937). **12**, 227–231 (1895)
11. Brock, T.: Milestones in Microbiology. Prentice-Hall International (1961)
12. Evans, A.: Causation and disease: a chronological journey. The Thomas Parran Lecture. Am. J. Epidemiol. **108**, 357–361 (1978)
13. Petrick, G.: The Arbiters of Taste: Producers, Consumers, and the Industrialization of Taste in the United States, 1900–1960. University of Delaware (2007)
14. Zeide, A.: In Cans We Trust: Food, Consumers, and Scientific Expertise in Twentieth-Century America. University of Wisconsin-Madison (2014)
15. Underwood, W.L.: Incidents in the canning industry of New England. In: Judge, A. (ed.) A History of the Canning Industry by Its Most Prominent Men, pp. 12–14. The Canning Trade (1914)
16. Goldblith, S.A.: Pioneers in Food Science: Samuel Cate Prescott, M.I.T. Dean and Pioneer Food Technologist. Food & Nutrition Press, Trumbull (1993)
17. Goldblith, S.A.: Japanese scientists and the POW's. Science. **104**, 302–303 (1947)
18. Goldblith, S.A.: A simply constructed biophotometer. Science. **104**, 449 (1946)
19. Richards, E.H.: The Chemistry of Cooking and Cleaning: A Manual for Housekeepers. Estes & Lauriat90 (1882)
20. Scloss, C.: Preserving the light: the photography of Wm. Lyman Underwood and Loring Underwood. In: Lyons, R. (ed.) Gentleman Photographers: The Work of Loring Underwood and Wm. Lyman Underwood, pp. 11–30. Solio (1987)

21. Prescott, S.C., Underwood, W.L.: Contributions to our knowledge of micro-organisms and sterilizing processes in the canning industry. Science. **VI**, 800–802 (1897)
22. Deming, O.L. (ed.): Science and Experiment as Applied to Canning. Sprague Canning Machinery Co (1902)
23. Pearson, G.S.: The Democratization of Food: Tin Cans and the Growth of the American Food Processing Industry, 1810–1940. Lehigh University (2016)
24. Prescott, S.C., Proctor, B.E.: Food Technology. McGraw-Hill Book Company, Inc. (1937)
25. Lueck, R.: Dr. S.C. Prescott. His contributions to the canning industry. Cann. Trade. **67**, 10–11 (1954)
26. Schatzberg, E.: Technology: Critical History of a Concept. The University of Chicago Press (2018)
27. Richards, E.H.: Euthenics, The Science of Controllable Environment: A Plea for Better Living Conditions as a First Step Toward Higher Human Efficiency. Whitcomb & Barrows (1910)
28. Bush, V.: Science, The Endless Frontier. Princeton University Press (1945). https://doi.org/10.1515/9780691201658-002

Chapter 9
Botulism: The Applied Science of Canning

Abstract America from 1919 to the mid-1920s. Well-publicized cases of botulism from canned foods threaten the entire industry. Scientists from the National Canners Association collaborate with State and Federal governments and with university researchers in an organized research program to solve the problem. Science is used collaboratively by industry and government to guide regulation.

9.1 Food That Kills

On December 14, 1895, the *Fanfare Royale Les Amis Réunis* performed at a funeral in Ellezelles, Belgium. Afterward, the musicians ate lunch at a local restaurant, "Le Rustic," where some of them had ham. Within a couple of days, more than ten men and boys had become dangerously ill and, before Christmas, three were dead. A local doctor attending the victims sent samples to the professor of microbiology at the University of Ghent, Emile-Pierre Van Ermengem [1].

At the time, the idea that food could transmit disease was not generally understood. Many still believed illness from food was caused by toxins known as "ptomaines," thought to form during the decay of protein-rich foods [2]. Van Ermengem, however, was the very model of a modern microbiologist. He applied methods similar to those that Russell and Prescott were using at the time to understand spoilage in canned food (Chap. 8) [3]. He took samples from the bodies of the deceased musicians and from the leftover ham and in both cases found similar rod-shaped bacteria which grew only in the absence of air. When he fed samples to animals, he saw they developed symptoms similar to those the band members had suffered. Critically, when he separated the bacteria from the ham juice with a fine porcelain filter, the juice remained deadly—suggesting a toxin produced by the bacteria and not the bacteria themselves was responsible. Van Ermengem named the organism *Bacillus botulinus*—*Bacillus* because their cells were rod-shaped and *botulinus* from the Latin for sausage, because the same symptoms had been frequently reported in people consuming spoiled sausage. In 1924, they were reclassified as *Clostridium botulinum*.

J. Coupland, *Containing Nature*, https://doi.org/10.1007/978-3-032-11882-0_9

Clostridium botulinum's preferred home is in the soil, where the oxygen-free (anaerobic) environment and moderate temperatures provide it an ideal ecological niche [4]. However, like many soil bacteria, they can easily find their way into foods. In most cases, this isn't a problem, because the conditions in the foods do not support their growth. Importantly, *Clostridium botulinum* cannot grow under acidic conditions, meaning most fruits are safe while meat and most vegetables are not. Even if the food is not acidic, *Clostridium botulinum* cannot grow in the presence of oxygen, and as most foods are exposed to air, it isn't a problem. But if the conditions are right and it does grow, *Clostridium botulinum* spoils the food by producing gases and a characteristic aroma. Much more seriously, however, it also produces the most toxic substance known.

Botulinum toxin is best known today by its brand name Botox®, and tiny amounts are used in cosmetic procedures to relax muscles in the face, causing unwanted lines and wrinkles to fade. However, the muscle-relaxing effects of only a few nanograms of botulinum toxin in food can easily lead to paralysis and death. The doctor's notes on Angel Deltenre, a 15-year-old boy who had been a member of the band from Ellezelles, provide a typical case: "December 19th: In the morning I was called in. No bowel movement and no vomiting since eating the ham. No pain. Pupils strongly dilated, partial paralysis of the eyelids. Breathing free, temperature normal, pulse 80 per min. Considerable weakness in the muscles, consciousness not affected. Purgative administered by enema had no effect. Called out again in the evening; pulse 100 per min, temperature 37.5 °C [99.5 °F]. Very rapid breathing. Patient could swallow almost nothing: everything coming back through the nose. He was now agitated; could not sit or lie comfortably. Lost consciousness briefly. Died 2 h later, fully conscious" [1]. Moderate cooking is sufficient to destroy the botulinum toxin, but the risk from undercooked or raw foods contaminated with *Clostridium botulinum* is very serious.

Botulism has been known since people first started preserving foods and unknowingly creating environments where the bacteria responsible can thrive [5]. Sausages were recognized as particularly risky, as air incorporated during their manufacture is consumed by residual enzymes in the meat. If the sausage is large and densely packed, then the center becomes anaerobic, and if there is no lactic acid fermentation to make the meat acidic, then *Clostridium botulinum* can begin to grow. The large blood sausages of Southern Germany were particularly hazardous, and it was from them the toxin was first isolated in the 1820s.

Perhaps the pig that went on to become the ham that was served in Ellezelles had rooted round in dirt contaminated with *Clostridium botulinum*. The bacteria had persisted—and perhaps even grown—in the low-acid and anaerobic brine barrel ever since the animal had been slaughtered and processed that summer. Because the ham was not cooked before serving, the toxin was active and caused the sickness and deaths.

The development of canning provided a new ecological niche for *Clostridium botulinum*. Sealed cans typically lack oxygen, and if the food inside isn't sufficiently acidic, they provide an ideal environment for the bacteria to grow and produce its lethal toxin.

Even more seriously in this new context, *Clostridium botulinum* has the capacity to form spores—dormant structures that allow some bacteria to survive periods of intense environmental stress. In soil, *Clostridium botulinum* cells would be killed by prolonged dry periods, but their spores survive until moisture returns. Unfortunately for us, these spores are also much more resistant to heat than the growing (vegetative) cells. While both *Clostridium botulinum* cells and the botulinum toxin can be destroyed by simple cooking, the spores are highly heat resistant. If even one spore survives inside the cooked can, it can germinate to form a viable cell which can then reproduce, produce toxin, and kill people. The question was: how can you be sure your cooking process was adequate? Nicolas Appert had claimed to know the answer.

9.2 Learning How to Cook Cans from Experience

In his 1810 *Book for All Households*, Nicolas Appert offered guidance on how long bottles of different types of food needed to be boiled. Acid fruits and asparagus merely required the canner to simply "bring the bath to boiling," while spinach needed a quarter of an hour, roast meat and vegetable soups half an hour, and various types of beans 1–2 h.

At first glance, this seems to tie in well with modern understanding: acid foods need less time than nonacid foods because only the latter can support the growth of *Clostridium botulinum*. On further consideration, however, it appears Appert was more concerned with food quality than safety. His way of checking his cooking was sufficient was to simply watch the bottles for a few days to make sure they didn't explode. The cooking times he selected were long enough for bottles not to ferment in the days following preparation, yet not so long that foods were overcooked and unappealing. Neither spinach nor beans are acid, yet Appert recommends a far shorter time for the delicate spinach leaves than the robust beans.

What Appert meant by his recommended cooking times were simply the times that worked for him. Indeed, when historian and food scientist Norman Cowell made calculations based on Appert's recommendations, many processes amounted to only a very mild pasteurization that might kill some bacteria but would be unlikely to damage any *Clostridium botulinum* spores present [6]. Indeed, at the boiling point of water, the destruction of *Clostridium botulinum* spores would be so slow as to be practically impossible. Appert's success may simply have been a combination of good luck, the low expectations of his customers, and perhaps the fact that inadequately processed bottles simply exploded from the buildup of fermentation gases before they could be sold and eaten.

The development of first brine baths and later pressure retorts for cooking cans allowed higher temperatures, making proper control of spores possible. These innovations also gave canners the ability to refine their processes by adjusting both cooking temperature and time. This increased flexibility enabled the development of time-temperature processing conditions tailored to different foods. However, the methods used to determine these conditions had not advanced much beyond those of Appert. Writing in 1927, the pioneering food technologist Olin Ball described how process development worked: "…a limited number of cans of the food were given the minimum process that was thought sufficient to sterilize. The cans were then held in storage, and if they did not spoil, the process that had been used was adopted as the proper process. If a considerable number of cans spoiled a more severe process was tried in the same way. This procedure was repeated as often as was necessary to determine a product that was sterile and that had satisfactory appearance and taste for a commercial product" [7].

Once a method was developed to everyone's satisfaction, it could be used until there were problems—a slight change to the recipe might lead to an outbreak of swollen cans or complaints that your food is mushy and overcooked—at which point the search had to start again.

Empirical knowledge like this, hard-won through trial-and-error, was valuable to those that held it. If a company knew how to process cans more effectively than their competitors, then they had a competitive advantage. In the early days, the only way to learn how to can foods was to figure it out for yourself. Later however, those setting up a new cannery could seek out a processor with experience elsewhere who could direct their cooking operations. Andrew Hapgood ran the salmon processing room for the Hume brothers in Sacramento (Chap. 5), while the Landreth's in Wisconsin had hired Francis Patterson to help them can their peas (Chap. 8).

These processors knew that their real value was not in their physical skill moving cans in and out of cookers, but their knowledge of how different foods could be cooked. They defended their positions by keeping their methods to themselves—and out of the hands of the cannery owner. Writing in 1937, Earl Chapin May recalls the power and authority of a processor in a cannery around the time of the American Civil War: "Each processor, theoretically, possessed his own secrets. Each was the sole custodian of an unparalleled 'formula'. Each, alone, knew a certain combination of time, temperature, and equipment which, when used at his sole direction, produced the world's finest most reliable canned foods" [8].

May goes on to describe the theatrics used by some processors to preserve their unique authority: "A processor in a canned crab plant in Hampton, Virginia would ostentatiously produce a vial of blue liquid from a safe and carefully add a couple of drops to the processing water before locking it away again from prying eyes. What was this magical ingredient that was essential to producing the best canned crab? Colored water." The Wonderful Wizard of Oz could not have conjured more convincingly from behind his curtain.

The first blow against the processors and their secret knowledge was the development of large-scale industrial equipment, particularly pressure retorts. If a canner was going to invest in an expensive piece of equipment to process cans, they would reasonably expect instructions on how use it effectively. While a canner or a processor wanted to keep their knowledge to themselves either to secure a commercial advantage or higher wages, an equipment manufacturer had different priorities. They wanted their machines to be used effectively by as many people as possible, so they began sharing information through publications and by organizing regional trade associations.

The second blow against the processors and their secret knowledge was the growing realization that science could offer more reliable answers than experience alone. Recall that scientists had proved largely unhelpful to canners for most of the history of the industry (Chap. 7), but this began to change in the 1890s when a new generation of microbiologists at American universities began to solve the practical problems of local canners (Chap. 8). Beyond simply solving the immediate local issue, the scientific approach offered distinct advantages. The scientists had a set of established methods (e.g., microscopy, chemical stains, and special types of selective nutrient media) and knowledge (e.g., the properties of different types of microorganism) that could be quickly applied to problems in canning. They had a network of scientific journals distinct from the canners' network of trade associations. Microbiology was advancing rapidly, and what was learned in one application could quickly be applied to others. For example, canners often described the canning defects they saw in terms of the "symptoms" they manifested. A "flat sour" referred to a can that tasted sour but showed no gas production, a "swell" bulged at the ends due to gas buildup, while a "springer" was usually caused by a sealing defect—the lid neither bulged nor curved inward but would spring back and forth when pressed. When they used their terms, their colleagues would know something about the problems they face. Microbiologists, in contrast, talked about the types and numbers of bacteria present.

By the start of the twentieth century, the pioneering work of Russell, Prescott, and others had shown a path toward a full scientific theory of canning. However, doing the work needed to make that happen would require a sustained effort, and that was going to be expensive. A single company didn't have the resources for the task, and even if they did, they would be tempted to keep their discovery a secret and gain commercial advantage from it. What was needed was an organization with deep pockets and an interest in the improvement of canning across the whole industry. They also needed a crisis to force them to act.

9.3 A Problem That Could Be Solved Becomes a Problem That Must Be Solved[1]

At the start of the twentieth century, there were good reasons to be suspicious of the food you bought at stores or in restaurants. The new industrial system of packaged foods made in distant, smoky factories was supplying a growing proportion of the American diet. The sanitary conditions were, by our standards, appalling, and there was widespread adulteration with undeclared and possibly dangerous preservatives and colors.

---

[1] Questions of the safety of canned food and the response of the canning industry in this period are the topic of Chapter 1 of Gabriella Petrick's PhD dissertation [9] and Chapter 3 of Anna Zeide's [10]. Much of this section draws heavily on their work, as well as Chapter 3 of Zeide's book [11].

Canned food, more than anything else, was emblematic of this new food system and the suspicion that went with it. The metal cans were hard industrial objects, very different from the traditional foods prepared and preserved in kitchens. The metal hid the food from the consumer, as historian Gabrielle Petrick described it: "consumers could not peer into the can … to see what the food looked like. Nor could they smell or taste the food before they purchased it" [12]. People were forced to trust the canners and the canners depended on people's trust.

Ironically, when done properly, canning was one of the safest methods of food preservation available at the time. The heat treatment destroyed harmful bacteria ubiquitous in fresh foods prepared under poor sanitary conditions, and there was no need for chemical preservatives. However, when cases of botulism did occur, the sudden and tragic deaths attracted national attention.

In one famous case from November 1913, a sorority party at Stanford University served a salad made from commercially canned as well as home-canned string beans [13]. Several young women contracted botulism, and one died. One of the remaining cans of the home-prepared beans was later described as "'blown' somewhat and its contents fermented and slightly sour," but the acid in the salad dressing may have masked any butyric taste from spoilage. Even if it was the home-produced string beans that were responsible, the outbreak was blamed on "canned food" and the reputation of the canning industry suffered.[2]

The movement toward federal regulation of food at the start of the twentieth century had led to several of the regional canning trade associations merging to form the National Canners Association (NCA)[3] in 1907 [10]. The NCA had resources and an interest in promoting the consumption of canned food nationwide. They recognized that if any canned food caused illness, consumers might decide "canned food" as a category was dangerous, and all manufacturers would suffer—regardless of who was actually responsible.

The NCA's first reaction was to defend the safety of canned food with a publicity campaign making use of some rather self-serving epidemiological data. "Only" about one can in a billion was associated with a botulism outbreak, and botulism "only" killed about 25 people each year, while 1700 died from other forms of accidental poisoning and 400 from lightning strikes [14]. This may have been true, but the idea that "this meal has a one in a billion chance of killing my family" is perhaps not one you would want to be top-of-mind for a consumer opening a can of food.

[2] Preservation of low-acid foods by canning in glass jars was a widely used method of home food preservation at the time. The recommended process for home canners at the time was to heat the sealed jars in boiling water on three successive days to allow any surviving spores to germinate on cooling and be killed by the next heating step. However, this process was laborious and frequently not completed properly. Best practice today requires the use of a pressure cooker, and the process remains hazardous unless the guidelines are followed strictly.

[3] The National Canners Association is the first national food industry trade group. It changed names to the National Food Processors Association in 1978 and Food Products Association in 2005. In 2007, it merged with the Grocery Manufacturers Association and in 2020 became known as the Consumer Brands Association.

Real action was needed to reduce the number of poisonings and, more cynically, to build public confidence by being seen to do something.

What focused their efforts was a flurry of cases in olives [15]. In many ways, canned olives were an ideal environment for *Clostridium botulinum*. The fruit were grown in California where spores were endemic in the soil, they are not acidic, and some were packed in glass (so the consumer could see them) and subjected to a lower heat than foods packed in metal cans (often only 30 min in boiling water). They were also consumed raw, without a cooking step that might have destroyed the toxin.

In August 1919, a party was held at a country club near Canton, Ohio, to celebrate the triumphal return of a war hero [10, 16]. The local dignitaries at the top table were served a special plate of "ripe olives, chocolate candy, Newport creams, and candied almonds." Within 3 days, seven people, including the war hero, were dead and botulism from the olives was to blame. The same batch of olives killed five people in Detroit and seven in Memphis.

Six months later, Mary Delbine prepared a meal for her family in the Bronx using jars of giant olives and died the following day. Three days after that tragedy, her bereaved family gathered for the meal, ate a salad prepared with the olives, and two of her sons, her husband, and an uncle also died. The reputation of the olive canning industry was in tatters, and the whole canned food industry was suffering by association. The need for action was pressing.

In late 1919, the NCA was involved in forming the Botulism Commission in California to coordinate the research work. Substantial funding was provided by the NCA, the Canners League of California, and the California Olive Association, but the leadership was from government and academia: Earnest Dickson, a Stanford microbiologist who had been studying botulism since the 1913 sorority outbreak at his university; Karl Friedrich Meyer, a distinguished Swiss-American bacteriologist and pathologist from UC Berkeley; and Jacob Geiger, an epidemiologist who had worked for the US Public Health Service at Camp Pike, Arkansas, the epicenter of the 1919 Spanish flu epidemic [15, 17, 18]. The chief of the USDA's Bureau of Chemistry (the forerunner of the FDA) ordered local staff to support the commission as far as possible without compromising their regulatory mission. The stakeholders were motivated, the funds were available, and it was finally possible to get the problem of processing solved.

9.4 Killing Spores with Heat

In 1913, the NCA set up their own laboratory in Washington DC, and their first chief of chemistry was Willard Bigelow. Bigelow had previously taught chemistry at the University of Oregon and in Washington DC public schools before he was recruited to the Bureau of Chemistry in USDA under its campaigning chief, Harvey Washington Wiley [19, 20]. At the USDA, Bigelow had been a key lieutenant in Wiley's campaign for federal oversight of food manufacturing and had provided essential analytical expertise in the campaign against dangerous additives and adulterants (see Chap. 10).

Bigelow's scientific background and his professional connections with government and industry enabled him to manage the scientific work required to tackle the botulism issue. His approach was in two parts—the first was conducted largely by microbiologists concerned with how *Clostridium botulinum* could be killed by heat and the second largely by physicists and engineers who measured and modeled the penetration of heat into a sealed can during cooking.

They focused their attention exclusively on *Clostridium botulinum* spores, because they were the most heat-resistant anaerobic pathogen known.[4] If they could identify the minimum conditions needed to destroy them, then they could be confident any other microorganisms present would already be dead. Studying *Clostridium botulinum* was difficult though, not only because it is so dangerous but also because it is an obligate anaerobe—it is poisoned in the presence of oxygen. Therefore, all the experiments had to be conducted in a carefully controlled, oxygen-free atmosphere. The first of the challenges was simply isolating the samples to be studied.

An environmental or clinical sample always contains a wide variety of different types of bacteria, making it difficult to determine which one is responsible for a particular issue. The nineteenth-century pioneers of microbiology—Koch, Pasteur, and others—developed methods to isolate a single type of bacterium from such mixtures including a technique called "streak plating." A suitably diluted sample would be spread across the surface of an agar plate so that the individual cells were separated. After incubation, each isolated cell would multiply into a visible colony—a small "blob" on the agar surface. Since each colony arose from the identical descendants of a single cell, it contained only one type of bacterium. The colony could then be carefully transferred to a sterile medium to create a pure culture. These were the methods Russell and Prescott had applied in their pioneering work on spoilage organisms in cans during the 1890s (Chap. 8). However, because the plates were grown in air, a variation of the method had to be developed for strictly anaerobic organisms like *Clostridium botulinum*.

The researchers took advantage of *Clostridium botulinum*'s unique ability to resist heat. They would begin by boiling a sample from food or a patient for a few minutes to kill all non-spore-forming organisms. The remaining material was then diluted in hot, liquid, agar and poured into long cylinders contained in glass tubes and capped with a disk of wax. The length of the cylinders prevented air from diffusing in, so only spores of anaerobic organisms could grow. Any visible colonies which formed gas bubbles in the agar gel, another characteristic feature of *Clostridium botulinum*, could be carefully fished out with a sterile wire loop and grown again as a pure culture. The cells were identified, first provisionally from

[4]There are other anaerobic bacteria that produce spores even more heat resistant than those of *Clostridium botulinum*. However, while they may spoil canned food, they do not produce a dangerous toxin. Other bacteria are important in acidic foods where *Clostridium botulinum* cannot grow, but this effort was focused on what came to be known as "low-acid foods."

their rodlike shapes and then, more conclusively, by their lethal effects on guinea pigs [21].

Having isolated their organism, the researchers could now ask the question they were interested in: how much heat was needed to kill it? To do this, they had to place spores in an environment similar to that found in the inside of a can—anaerobic and surrounded by the components of a low-acid food—and then expose them to a given temperature for a given time. Finally, they had to determine if any spores had survived this heating process. They could then repeat the experiment to find the minimum time needed to destroy *Clostridium botulinum* spores at each temperature.

Because they needed to give each sample of spores a precisely specified treatment—this temperature for that time—they knew they couldn't do the experiments directly in cans of food. Cans are large, so when you cook them, the center warms up more slowly than the outside, and at the end of the experiment, spores in different parts of the can would have been exposed to very different amounts of heat. Instead, the samples used for the microbiology studies had to be very small, so heat conduction into the center was so fast that it could be ignored, and all the spores received effectively the same heat treatment.

Bigelow and his team did this by first encouraging *Clostridium botulinum* cells to form spores by gently heating them and then sealing the spores into thin glass tubes with some deoxygenated liquid medium—frequently the juice from corn, olives, or whichever food they thought most relevant. The tubes were small, typically with a diameter of only 7 mm and a wall thickness of 1 mm, so, when they were plunged into hot oil, the scientists could be confident the contents were all exposed to the temperature of the oil for the time they were immersed. They then cooled the tubes in ice, broke them open, and tried to grow the contents in a nutrient agar. If anything had survived the heating and was able to grow, then that heat treatment was inadequate.

One of the first major studies of this kind was published in 1920 by Bigelow and James Esty, a young microbiologist who had joined the NCA from Brown University only the year before, describing a range of unidentified organisms found in canned foods [22, 23]. Two years later, Esty returned with the definitive study on *Clostridium botulinum* itself [24]. Unsurprisingly, it took much longer to sterilize the tubes at lower temperatures. For example, in one study, it took 4 min to sterilize tubes of spores at 120 °C, 33 min at 110 °C, and 330 min at 100 °C. However, an interesting pattern emerged from the data: increasing the temperature by a certain amount, for one important variety of *Clostridium botulinum* about 10 °C, reduced the time needed to sterilize the tubes by a factor of 10 [25, 26]. This relationship would later provide a useful tool to calculate sterilization times at different temperatures.

Also present in this early data was an apparently trivial finding that was misunderstood at the time and whose implications would be neglected until the 1940s. Given the effort to ensure that all spores were of the same type and exposed to the same temperature, it seemed logical to expect them to die at the same time—but they didn't.

What's more, there was an unusual problem with error in the thermal death measurements themselves. All scientific measurements have some level of random noise due to the small variations in materials or technique, but the error Esty saw was not random. At either high temperatures and long times or at low temperatures and short times, the results were consistent: the tubes were consistently sterile or not. However, in intermediate cases, the results were inconsistent with different tubes giving different results, a phenomenon he called "skip-stops." When Esty repeated the experiments, not with a single tube for each treatment but with a larger number, perhaps 40, he saw the pattern [27]. In this intermediate range, the proportion of the tubes got greater the longer the sample was heated for. What is more, the pattern was logarithmic—increasing the heating time by a certain amount decreased the proportion surviving tenfold.

However, these concerns were not pursued by the NCA researchers in the 1920s. What they did reflects their practical concern about the "worst-case scenario" for canning. They had already selected the most heat-resistant spores they could find for their study. So, when they realized that a higher number of spores required more time to kill, they tested the maximum number they could fit into a tube—about 60 billion. This was vastly higher than anything that could exist in nature—the glass tube was packed with spores so tightly that they were virtually touching one another. It was the "worst case" they could generate, and these were the numbers Esty reported.

9.5 The Temperature Inside a Can

It is obvious to anyone who has ever been in a kitchen that the center of a piece of food cooks more slowly than the surface. The studies on thermal death times of *Clostridium botulinum* spores had purposely used thin glass tubes to be sure the samples were all exposed to the same temperature for the same time, but the contents of a can heat up slowly and unevenly. Prescott and Underwood had been aware of this in the 1890s; however, when they tried to study heat penetration inside a can of corn, the only measurements they had were from special "registering thermometers" with a kink in the tube to trap the mercury at its highest point and measure the maximum temperature reached [28]. They used these to show the center of a can of corn in a retort set at 118.8 °C (245.8 °F) only reached that temperature after 55 min. The second thread of study for the National Canners Association team in the 1920s involved trying to measure the evolution of temperature at the center of the can, and here physicists and engineers took the lead.

Willard Bigelow had been in his role as laboratory director of the National Canners Association for 6 years in 1919, and he was still looking for the talent he needed to solve the food safety problems faced by the industry. He met a promising candidate at an event held at the Chemist's Club in midtown Manhattan—a young man just back from the war in Europe, a young man whose mathematical approach would transform the science of canning.

Olin Ball was the son of a mechanic from Abilene, Kansas.[5] After finishing high school, he had served with the Kansas National Guard as part of General Pershing's military expedition to capture Pancho Villa in Mexico. He took a few college courses by correspondence, but before he could graduate, the Great War started, and Corporal Ball reenlisted. He served on the Western Front, again under Pershing, and rose to the rank of captain before he was discharged in New York. He was looking for a job, and his meeting with Bigelow set the direction for his career.

Ball studied at Georgetown while working in the Washington Laboratory of the National Canners Association. He quickly completed his undergraduate degree and then enrolled in a master's program in chemistry with a focus in mathematics. A research thesis at an elite university like Georgetown required substantial scientific content, but Ball had the additional challenge of doing this while retaining enough practical applicability that would interest his employer. His 1922 thesis, *The Electron Theory Applied to Themo-Electricity*, was seen as valuable for both groups. It was about thermocouples.

What the NCA needed was a way to measure the temperature in the center of a can of food while it was being cooked in a commercial pressure retort. Some early efforts had attempted to fit a glass thermometer, so the bulb rested in the center of the can while the temperature scale was visible outside, but this unsatisfactory as the glass of the thermometer could conduct heat into the food and also because it was not well suited to commercial retorts where many cans are frequently stacked up and sealed at in high pressure. It would be better if the thermometer could be read with a thin, flexible wire—a thermocouple.

Various workers throughout the nineteenth century had noticed the electrical properties of certain materials depended on temperature and could, in theory, be used as thermometers [31]. However, creating something suitable for industrial use proved challenging and was only beginning to be addressed by the end of the century. Once thermocouples became available in the early twentieth century, canners were quick to exploit them to solve their heat measurement problems. There were some early studies on heat penetration into cans in the 1910s using thermocouples, but the first systematic work was completed in the laboratories of the National Canners Association and reported at the end of that decade [32].

Bigelow and Ball's NCA group collaborated with the Philadelphia instrument manufacturers Leeds and Northrup to develop a thermocouple sufficiently robust to work in commercial canning retorts [32]. A thin copper-constantan electrode was carefully insulated to prevent heat conduction and threaded into the center of the can through a hole drilled in the lid and fixed in place with a thick rubber gasket that could provide a good seal against the steam pressure. Multiple cans, each containing their own thermocouple, could be placed inside a commercial retort. Their wires were connected to a voltmeter outside the retort through a second pressure sealer known as a "stuffing box." Once they had established their instrument, they used it

[5] The information on Ball's life is largely taken from Richard Stier's biography [29] and also from Lori Valigra's profile [30] of him.

to build a systematic understanding of the factors that affected heat penetration into canned foods.

The data was recorded by hand as a table of the temperatures measured at different times which were typically then converted to graphs, heat penetration curves, with temperature on the vertical axis and time on horizontal. The graphs started with a flat lag phase because the temperature of the center did not begin to increase until heat was transferred from the steam of the retort, through the can and its contents, to the thermocouple in the center. Once the core temperature began to change, it increased sigmoidally and only slowly approached the temperature of the retort. Cooling at the end of the process, after the retort was vented, was rather quicker than heating but showed a similar lag before a sigmoidal decrease.

The NCA team, along with researchers at the USDA, systematically examined how various factors affected heating curves. They found that the size of the can mattered—larger cans took longer for heat to reach the center. The initial temperature of the food also played a role: if the food entered the can hot, it required less time in the retort. Likewise, higher retort temperatures led to faster heating at the center of the can.

They also studied the effects of the can contents on heating profiles but found few general rules that could predict the behavior of all foods. Heat conduction through a solid is relatively slow, while convection currents in a liquid help mix the contents and allow it to heat more quickly. As a result, the center of a can of soup would be expected to reach the target temperature faster than the center of a can of salmon. Other foods were more complicated, with solid parts and liquid parts interacting with one another and heating at different rates. For example, tomatoes in cans appeared to move around and stir the brine, speeding up heat transfer, while spinach leaves stuck together preventing effective mixing and so were slow to heat.

Heating and cooling curves depended on so many factors that they had to be measured on a case-by-case basis. The developments in thermocouple technology led by Ball and Bigelow made this task straightforward, and experimentally measuring heat penetration curves remains an essential part of thermal process validation to this day.

9.6 Calculating and Legislating Safety

The microbiology measurements of how quickly *Clostridium botulinum* spores were killed in thin glass tubes at different temperatures along with the empirical data on how different cans heated and cooled provided the two pieces of information Bigelow needed to answer the question everyone cared about: is this process adequate? [32] He recognized spores weren't only being killed when the center of the can was at the maximum temperature—some would be killed while the food was being heated, others during cooled. To assess the effectiveness of a process, all of this had to be taken into account.

In essence, his approach was as follows: if you knew spores were destroyed in 10 min at 115 °C, then you could use the *z*-value, approximately 10 °C, to calculate how long it would take at any other temperature: 100 min at 105 °C or 1 min at 125 °C. These sterilization times could be easily converted into sterilization rates: a 10 min thermal death time meant a sterilization rate of 0.1 per min. That is to say, you would achieve 10% of complete sterilization each minute at 115 °C, 1% each minute at 105 °C, or 100%—complete sterilization—in a minute at 125 °C.

Next take the time-temperature graph measured using thermocouples at the center of the can and change the y-axis from temperature to sterilization rate. The curve starts at zero when the center of the can was too cool to have any lethality at all—it rises to a maximum when the core temperature is at its hottest and then drops back to zero as the can cools. The area under this graph represents the total thermal destruction done to the spores at the center of the can. If it exceeds one, then you can be confident the spores have been destroyed, and the process is adequate. If it is less than one, then either the temperature or the time would have to be increased, another heat penetration curve measured, and the process repeated [32]. In the age of computers, this sort of task would be trivial, but it was laborious without them.

It was Olin Ball who came up with a simpler approach.[6] He saw that if you plotted the heat penetration curves on semilogarithmic graph paper, they often appeared to be straight lines. You could then simply take the equations of the lines for heating and for cooling along with the lag time before the can center began to heat, put them into his formula, and calculate lethality. By the middle of the 1920s, Ball's formula method was established and freely available in a variety of more convenient forms. All a canner needed to do was measure a time-temperature curve for their particular food and follow the formula, and they could be certain their process was adequate [7, 34].

"Adequate" is an important term here. The process selected might be adequate yet completely unsuitable to make marketable food. Perhaps some beans would be undercooked, or spinach overcooked, or perhaps all the vitamins had been destroyed, or perhaps the food still spoiled due to the presence of bacterial spores more heat-resistant than *Clostridium botulinum*. "Adequate" meant a process was sufficient to kill spores according to the theories put together by the NCA researchers. So long as that was achieved, a canner could adjust the conditions as they saw fit to make the best product possible.

In addition, because "adequate" was clearly defined, it could also be legally enforceable. Under the terms of the Pure Food Act, the government could only remove foods from the market if they could be shown to be adulterated. However, even given the number of outbreaks in the early 1920s, the NCA had been correct to claim botulism from canned foods was incredibly rare. It would be impossible for USDA investigators to identify cans containing *Clostridium botulinum* and remove

[6]While Ball's ideas were straightforward, his presentation frequently was not. Historian of technology Colleen Dunlavy quotes from her interview with one of his colleagues: "a brilliant and outstanding scientist, but he could bever be content to use the standard symbols that are used in physics… [H]e invented his own every time" [33].

them before there was an outbreak—you would have to open each one to check. However, the new standards allowed a clear determination of whether a process was adequate or not—even if an individual processed can contained no *Clostridium botulinum*. This determination formed the basis for new regulations.

This was brought into force in the State of California in May 1925 in the form of a law requiring any commercial cannery using a pressure cooker be equipped to "carry out such rules and regulations as the state board of health may adopt for the sterilization of such agricultural food products" [35]. These regulations included the process verification rules established by NCA research, but also many nonscientifically based, but eminently sensible, practices were established from industry experience. Companies had to ensure uncooked cans could not be mixed with cooked, that the food received was wholesome and was cooked promptly, that the apparatus was well maintained and reliable and the thermometers were present and calibrated, that staff operating the retorts were trained, and that good records were kept. All of these were common in the more responsible parts of the industry, but now everyone had to meet the standards. The 1925 Act also created a cannery inspection service paid for by a tax on the empty cans bought by canners (initially 20 cents per thousand cans) collected by their trade associations. The canners were willing to pay for their own regulation because of the dreadful costs to their reputations of having poor quality and dangerous canned goods on the market.

As Olin Ball wrote soon after establishing the first practical method: "It is hoped that the facts that have been set forth will be of material assistance to the canner in the search for fundamental causes of many of his troubles, and in the effort to solve many of his problems of processing; or, that they may at least give to others a few pointers which will enable them to make more rapid progress in future research" [34]. Ball's modest hopes were well-founded. While the models have been significantly improved since then, the NCA studies in the early 1920s remain the basis of modern canning science and regulation.

9.7 No Guarantees

The regulations worked. From 1926 to 1930, there were no cases of botulism reported from commercially canned foods, while home-canned foods—not covered by the Act—remained a problem [18]. But did the standards guarantee safety?

Recall James Esty's puzzling data from 1922. The minimum processing time increased with the number of spores in each tube, and, at intermediate times, only a fraction of tubes studied were sterilized. Esty pragmatically chose the worst-case scenario, defining the thermal death time as the time needed to sterilize every one of the sealed tubes when each packed was as densely as possible with spores—and then he moved on. Esty and his contemporaries speculated on possible reasons—perhaps the phenomenon was due to compounds released by injured cells? Perhaps some of the cells were protected by clumping? Perhaps there was some variability in the properties of the spores? But they never followed up in a systematic manner.

They shouldn't have been surprised for, even in 1922, this was not a new observation. As far back as 1910, the English microbiologist Harriet Chick had treated non-spore-forming bacteria with hot water [36]. She found that at each temperature, there was a characteristic heating time, often indicated with the symbol D, that would kill 90% of a given organism—a tenfold reduction. If you heated for twice that time, 2D, 99% would be killed (90% in the first half, then 90% of the remainder in the second), and 99.9% would killed after a period of 3D. Just as Esty would see years later with spores at higher temperatures, it was a logarithmic function.

Unlike Esty though, Chick offered the correct explanation for what was going on. She argued the way she knew if a bacteria remained alive was if it could reproduce and grow to form a visible colony of cells on an agar plate. Even though the processes of bacterial reproduction were largely unknown in her time, she hypothesized that it must depend on molecules, probably proteins, in the cell. The process of killing a bacterium by heat then is just the kinetics of the chemical reaction needed to destroy the most fragile of these proteins, and these reactions were known to proceed in the same logarithmic manner.

More critically, Chick also realized why the issue with the logarithm was so important: "it is no longer permissible to talk of the 'thermal death-point' of any particular species of bacteria" [36]. Imagine a case where 90% of the spores are killed in 10 min and a can contains initially 1000 spores. After 10 min, there are 100 left, after 20 min 10, and after 30 min only 1—what happens next? The logarithmic model suggests that after 40 min, there would be one-tenth of a live spore left. A reasonable interpretation of this apparently nonsensical conclusion is that each can has a 10% chance of containing a spore or that we would expect a surviving spore in one can out of every ten processed. This was the "skipping" Esty had seen at intermediate processing times, when a fraction of his thin tubes contained live bacteria.

By Chick's logic, increasing processing time could reduce the odds of a spore surviving but never guarantee it. But that is exactly what Esty had done, and his data was the basis for the scientific rules of canning. How safe was canned food?

Charles Stumbo grappled with this question in the 1940s while working at the central research laboratory of the Food Manufacturing Company in San Jose, California.[7] He reasoned that Etsy had measured his thermal death time as the time required to sterilize all ten thin glass tubes in a given group, when each tube had

[7] Corporate research labs were important in this period, and many of the important scientists would move between roles over the course of their career. The Food Machinery Company (soon after to be known as the Food Machinery and Chemical Company and currently the FMC Corporation) was formed in a merger of the Sprague-Sells Company, which made canning equipment, with another agricultural equipment company [33]. Stumbo worked for them briefly, before moving to an academic career at the University of Massachusetts. Olin Ball followed a similar career path, moving between industry and academia while pursuing similar research questions. After his work with the NCA, he moved to the industrial labs of the American Can Company in Illinois before finishing his career at Rutgers University [29].

been packed tightly with approximately 60 billion. What did this mean terms of D values? Stumbo argued that "thermal death time" meant the original 60,000,000,000 had to be reduced to 0.1—a 10% chance of a single spore surviving in any single tube so a likelihood that none of Esty's ten replicates showed growth. This corresponded to approximately 12 tenfold reductions, or a 12D process [37–40].

A 12D process was a lot, but even that was choice about what was acceptable rather than an absolute guarantee of safety. You could always reduce the risk by increasing the cooking time—but what level of reduction was "safe enough"?

The problem was you could never know exactly how safe your product was. While you could calculate the number of tenfold reductions your process achieved, you could never know the final number of spores remaining because you didn't know how many were present in the food at the start. For example, if you performed a 3D process on a sample that contained ten spores, you might expect one can in a hundred would have a spore at the end (i.e., ten divided by ten three times). On the other hand, if your initial sample was more heavily contaminated and had 10,000 spores at the start, then each can might have ten spores at the end. Again, this is what Esty has seen on the 1920s—tubes packed with more spores at the start needed longer to sterilize.

Stumbo didn't know the answer to these questions, but, based on his experience, he made estimates. He argued that only a tiny fraction of any organisms present in unprocessed food would be the heat-resistant *Clostridium botulinum* we need to worry about. He estimated that if a food contained a million organisms per gram, perhaps only ten of them would be *Clostridium botulinum* spores. What would it mean if you used Ball's method to validate a process for a can containing food with this level of contamination? In the critical volume at the center of the can, there might be 10 g of food—a hundred organisms. According to the rules worked out by Ball, those spores would be the ones to get the minimum heat treatment as specified by Esty's thermal death numbers. What happened to them would determine the safety of the product.

Applying that same 12D thermal treatment not to the 60 billion spores in Esty's glass tubes but to the 100 organisms Stumbo estimated might be present in the cold spot a can meant about one surviving spore in six billion cans—not zero risk, but a very, very small number. As Stumbo said: "Comparing this with the many hazards of our present day life, we realize that commercially canned foods as they are now processed cannot, by any stretch of the imagination, be considered a health hazard from the standpoint of their containing viable spores of *Cl. botulinum*" [39].

Stumbo's estimate was based on solid scientific reasoning from Esty's data and Chick's theories which he combined with his experience of typical prevalence of *Clostridium botulinum* and his judgment that, while no process could be absolutely safe, this was safe enough.

9.8 The Science and Technology of Safe Food

The flurry of work on botulism in the early 1920s was a breakthrough in the science of canning. In the immediate term, the results supported regulations and the regulations worked. The number of cases of botulism from commercially canned foods went to zero, at least for a few years, which was both a public health success and, for the canners, a public relations triumph. In the longer term, their achievements remain the basis for the regulation of thermal processing a hundred years later. However, it would be a mistake to see this as a purely scientific success, as both the conclusions of the scientists and the success of the regulations depended on empirical knowledge gained from practical experience in factories.

First, it is worth remembering that, even at the height of the botulism scares in the 1910s, the best canners had a pretty good idea how to safely process foods. As they had pointed out at the time, their canned foods were remarkably safe, and when a problem did arise, then home-canned foods or shoddy manufacturers were usually to blame. Even before starting the experiments, many canners realized the 30 min in boiling water used by some olive processors was an exceptionally mild treatment and likely problematic.

What the NCA ended up recommending was a much more conservative process and one that was claimed to be scientifically justified. However, as we have already seen, the concept of thermal death time the regulations depended upon was flawed. In addition, what was taken as a worst-case scenario—the tubes packed with as many spores as would fit—was not an absolute number but depended on how the experiment was set up. In Esty's original work, the glass tubes he used contained 2 ml of sample, about 40 drops. The tubes had to be thin to achieve rapid heating, but their length was simply an experimental convenience. If the tubes had been longer, they would have contained more spores, and the thermal death time measured would have been greater.

Despite this, the numbers Bigelow and Ball came up with from Esty's data seemed reasonable to the canners based on their practical experience. While not recorded in the literature, there must have been a series of "common sense" check on the scientific recommendations.

Second, it is worth remembering that while the scientifically based process calculations were important, at least some of the improvements in food safety achieved in the 1920s were due to legislation requiring all canners use best practices. Best practices, commonly referred to today as good manufacturing practices (GMPs), are simply the rules of operation that responsible processors have found from experience to be important. Some of these—like keeping canneries clean and only packing the most wholesome food—can help reduce the numbers of microorganisms going into the cans. According to Esty's understanding of a single thermal death time, this recommendation would have made no sense bacteriologically, as all the organisms would be destroyed anyway by heating. It took Charles Stumbo's work on the implications of logarithmic reductions in the 1940s to understand that if there were more spores at the start of the process, there would be a higher risk of some surviving at the end. Still, the "clean and wholesome" rule made perfect sense to the canners in 1925 and became part of the regulations. Other best practices required in 1925—making sure the equipment worked well, making sure the thermometers were properly calibrated, making sure all cans sold had been properly processed, and keeping good records of production—were simply "common sense" based on practical experience rather on any scientific investigation.

Interestingly, it is not the use of inadequate processes but a failure to maintain these best practices that is responsible for the rare modern outbreaks of botulism from commercially canned food [41]. Some examples include cases from Vichyssoise soup (1971, believed to be due to workers using the wrong retort), Alaskan salmon sold in England (1979 and 1982, believed to be due to improper sealing, meaning the cans became contaminated after cooking), and chili sauce (2007, due to an improperly maintained retort).

The logarithmic nature of thermal sterilization means that even a legally adequate process might leave a single *Clostridium botulinum* spore alive; Stumbo estimated this risk to be as low as one can in six billion. It would be possible to calculate scientifically how much lower this risk would be for a more intense process, but it would be misguided to think there would be to any resulting improvements in food safety. The risk of human error, someone failing to conduct the process as designed, is a far greater.

9.9 Reflections for Modern Scientists: Putting Science to Work

Russell and Prescott's successes in the 1890s depended on their personal relationships with the canners they worked with and willingness to learn and share knowledge (Chap. 8). Certainly, these attitudes of work were essential to the much greater successes of the NCA and the Botulism Commission in the early 1920s: Ball and Stumbo moved between industry and academic roles, Esty sampled canneries looking for strains of *Clostridium botulinum*, Ball developed simpler calculations that were useful in industry, and work was disseminated both in peer reviewed publication (for scientists) and NCA reports (for canners). What was different here was the scale and degree of coordination of the enterprise.

The NCA and the California canning groups had deep pockets and made strategic decisions about how their money should be spent to benefit their industries. The existential threat of the botulism outbreaks in olives forced them to take bold action. But, even if the canners had the economic means and capacity to address the botulism issue themselves, they wisely brought in partners at the start of the endeavor. The Botulinum Commission was put together including representatives from state and federal state government as well as from academia. Federal and state governments were interested in improving public health. They saw canning as an established and generally accepted method of food preservation that could be improved.

It didn't hurt that senior leadership at NCA, especially Bigelow, had good personal connections with the regulatory agencies (see Chap. 10 for more discussion on the relationship between NCA and USDA in this period). Academic researchers and institutions were expected to work on practical industrial problems. There were professors of medicine and microbiology deeply engaged in public health work, including work on botulism, and there had been about 25 years of academic research on canning since Russell and Prescott.

Putting together such a diverse group before starting the work led to greater acceptance of the final recommendations. As we have seen, the guarantee of safety was perhaps less secure than was implied at the time, but because all the relevant groups were part of the process, it was seen as acceptable.

We should pause before moving on from this story and reflect on who benefitted from this remarkable alignment of interests. In the 1920s, case all seems well: there were fewer cases of botulism, and the industry benefitted from the improved regulations. Probably some poorly capitalized or poorly managed canneries who could not meet the new standards were forced out of business; but it would take an extremely libertarian perspective to see this as a bad thing. Other cases, such as the collaboration to develop HACCP procedures in the 1960s led by Pilsbury and NASA, fit a similar pattern. However, there are other transformative changes, for example, genetic modification of crops or irradiation of foods, whose benefits are more seriously contested. In these cases, the cozy relationship between government agencies, industry, and academia is seen, by some, as against the public interest. In the next chapter, we will see how such a conflict developed in the history of canning around the issue of BPA.

References

1. Devriese, P.P.: On the discovery of Clostridium botulinum. J. Hist. Neurosci. **8**, 43–50 (1999)
2. Geist, E.: When ice cream was poisonous: adulteration, ptomaines, and bacteriology in the United States, 1850-1910. Bull. Hist. Med. **86**, 333–360 (2012)
3. Van Ermengem, E.: A new anaerobic bacillus and its relation to botulism. Rev. Infect. Dis. **1**, 701–719 (1979)
4. Lewis, K.H., Cassel, K.: Botulism. U. S. Department of Health, Education, and Welfare, Public Health Service (1964)
5. Erbguth, F.J.: Historical notes on botulism, Clostridium botulinum, botulinum toxin, and the idea of the therapeutic use of the toxin. Mov. Disord. **19**, 2–6 (2004)
6. Cowell, N.D.: An Investigation of Early Methods of Food Preservation by Heat. University of Reading (1994)
7. Ball, C.O.: Theory and practice in processing. Canner. **64**, 27–32 (1927)
8. May, E.C.: The Canning Clan. Macmillan Publishers Limited (1937)
9. Petrick, G.: The Arbiters of Taste: Producers, Consumers, and the Industrialization of Taste in the United States, 1900–1960. University of Delaware (2007)
10. Zeide, A.: In Cans We Trust: Food, Consumers, and Scientific Expertise in Twentieth-Century America. University of Wisconsin-Madison (2014)
11. Zeide, A.: Canned. University of California Press (2018)

12. Petrick, G.M.: An ambivalent diet: the industrialization of canning. OAH Mag. Hist. **24**, 35–38 (2010)
13. Wilbur, R.L., Ophuls, W.: Botulism – a report of food-poisoning apparently due to eating of canned string beans, with pathological report of a fatal case. Arch. Intern. Med. **14**, 589–604 (1914)
14. Anonymous: What every canner should know. Natl. Canners Assoc. Bull. **89-A**, 31 (1922)
15. Young, J.H.: Botulism and the ripe olive scare of 1919-1920. Bull. Hist. Med. **50**, 373–391 (1976)
16. Horowitz, B.Z.: The ripe olive scare and hotel Loch Maree tragedy: botulism under glass in the 1920's. Clin. Toxicol. **49**, 345–347 (2011)
17. Meyer, K.F., Steele, J.H.: Karl Friedrich Meyer. J. Infect. Dis. **129**, 3–10 (1974)
18. Meyer, K.F.: The protective measures of the state of California against botulism. J. Prev. Med. **5**, 261–293 (1931)
19. Anonymous: Dr. W. D. Bigelow, Research Chemist. The New York Times, 27 (1939)
20. Blum, D.: The Poison Squad. Penguin Press (2018)
21. Dickson, E.C., Burke, G.E.: Botulism: a method of isolating bacillus botulinus from infected materials. JAMA J. Am. Med. Assoc. **71**, 518–521 (1918)
22. Anonymous: J.R. Esty Dies; Director of the N.C.A. West Coast lab. Inf. Lett. National Canners Association. **1485**, 175–180 (1954)
23. Bigelow, W.D., Esty, J.R.: The thermal death point in relation to time of typical thermophilic organisms. J. Infcct. Dis. **27**, 602–617 (1920)
24. Esty, J.R., Meyer, K.F.: The heat resistance of the spores of B. botulinus and allied anaerobes. XI. J. Infect. Dis. **31**, 650–663 (1922)
25. Weiss, H.: The heat resistance of spores with special reference to the spores of B. botulinus. J. Infect. Dis. **28**, 70–92 (1921)
26. Bigelow, W.D.: The logarithmic nature of thermal death time curves. J. Infect. Dis. **29**, 528–536 (1921)
27. Esty, J.R., Williams, C.C.: Heat resistance studies: I. A new method for the determination of heat resistance of bacterial spores. J. Infect. Dis. **34**, 516–528 (1924)
28. Prescott, S.C., Underwood, W.L.: Contributions to our knowledge of micro-organisms and sterilizing processes in the canning industry. Science. **6**, 800–802 (1897)
29. Stier, R.F.: Dr. C. Olin Ball. In: Powers, J.J. (ed.) Pioneers in Food Science, vol. 2, pp. 79–112. Food and Nutrition Press (2005)
30. Valigra, L.: A pioneer in thermal death-time standards. Food Qual. **18**, 39–41 (2011)
31. Hunt, L.B.: The early history of the thermocouple. Plantin. Met. Rev. **8**, 23–29 (1964)
32. Bigelow, W.D., Bohart, G.S., Richardson, A.C., Ball, C.O.: Heat penetration in processing canned foods. Natl. Canners Assoc. Bull. **16–L**, 128 (1920)
33. Dunlavy, C.A.: Food Machinery Corporation's Central Research Department: A Case Study of Research and Development, 1942–1954. University of California (1980)
34. Ball, C.O.: Thermal process time for canned food. Bull. Natl. Res. Counc. **7**, 5–76 (1923)
35. Cannery Inspection Act. 931–933 (1925)
36. Chick, H.: The process of disinfection by chemical agencies and hot water. J. Hyg. (Lond). **10**, 237–286 (1910)
37. Stumbo, C.R.: Bacteriological considerations relating to process evaluation. Food Technol. **2**, 115–132 (1948)
38. Stumbo, C.R.: Further considerations relating to evaluation of thermal processes for foods. Food Technol. **3**, 126–131 (1949)
39. Stumbo, C.R.: Thermobacteriology as applied to food processing. Adv. Food Res. **2**, 47–115 (1949)
40. Stumbo, C.: Thermobacteriology in Food Processing. Academic Press (1965)
41. Pflug, I.J.: Science, practice, and human errors in controlling clostridium botulinum in heat-preserved food in hermetic containers. J. Food Prot. **73**, 993–1002 (2010)

Chapter 10
BPA: Conflicting Communities of Scientists

Abstract America at the turn of the 21st century. A community of environmental scientists identifies a novel risk from a component of can linings, and campaigns to get it banned. The limits of what we can know about risk. How the social networks of scientists shape their values and conclusions. Science and politics.

10.1 The Materials of Cans

In the early nineteenth century, the London canners improved Appert's food preservation methods by using new existing material—tinplate (see Chap. 3). Tinplate is made by coating a sheet of iron (later steel) with a thin layer of tin. The iron provides the necessary strength but is prone to rust, especially when in contact with acidic foods, which can spoil the contents and cause leaks. Tin resists corrosion but is too soft and costly to use alone. By combining the two, tinplate offered both durability and corrosion resistance at an affordable price. Although the production of tinplate was technically complex, it had already been refined by Welsh engineers, making it readily available for London entrepreneurs to use in food preservation.

Despite its advantages, tinplate had its drawbacks. One issue was the temptation to reduce costs by using thinner layers of tin which increased the risk of tiny defects—such as pinholes or scratches from handling—that exposed the underlying iron to corrosion. Additionally, although tin is less reactive than iron, it can still partially dissolve into certain foods, gradually thinning the protective layer and making corrosion more likely. Finally, tin can react with sulfur-containing foods to form unsightly black sulfide compounds.

To address these issues, canners quickly realized that an extra layer of protection was often necessary: a can coating. Early manufacturers commonly painted the exterior of their cans to prevent rust, but the interior required a more sophisticated solution. This inner coating had to withstand high cooking temperatures and prolonged contact with food, avoid reacting chemically to produce off-colors or flavors, and, crucially, be nontoxic.

J. Coupland, *Containing Nature*, https://doi.org/10.1007/978-3-032-11882-0_10

The earliest can coatings were created by baking layers of oil onto the tinplate surface, in a process similar to seasoning cast iron cookware. These coatings typically used drying oils such as linseed or tung oil, sometimes mixed with plant resins. Drying oils are also found in traditional wood varnishes and have chemical properties that allow them to quickly oxidize and form a waterproof film. Once applied to the tinplate, the oil was baked to create a thin solid film that effectively prevented direct contact between the food and the metal. However, this method had several drawbacks. The baking process was time-consuming, the coating was quite soft and easily damaged during cooking, and the oxidized oils sometimes imparted a rancid flavor to the food. A more effective solution emerged with the development of epoxy resins in another industry, offering a more durable and food-safe alternative.

When bisphenol A (BPA) was synthesized in a Russian laboratory in 1891, it was a completely new molecule—something that had never existed on Earth before. Initially, it was merely an interesting compound, but its significance grew in the interwar years when German chemists discovered how to link many BPA molecules together to form long chains, known as polymers. As the reaction proceeded and the length of the polymer chain increased, the samples became more viscous and eventually solidified to form a plastic. The plastics formed from the polymerization of BPA include polycarbonates—hard, clear, and very strong—and epoxy resins, commonly used as coatings and glues due to their rapid setting and excellent adhesive properties.

Commercial epoxy resins first became available shortly after the Second World War as protective coatings on industrial equipment and as dental adhesives, but they were soon adopted as liners for food cans [1]. Epoxy resin can liners offered immediate advantages: they are strong and flexible and bond tightly to metal surfaces. They also withstood the high temperatures of the retort and were compatible to a much wider range of foods than the drying oils they quickly replaced. Even better, they didn't require a lengthy baking process. By the end of the twentieth century perhaps 95% of cans had epoxy coatings. However, by then, there were widespread concerns that this miracle material was itself dangerous.

The story of the fight over BPA offers fresh insight into the complex interplay between science, politics, and technology. In earlier chapters, we saw scientists initially disconnected from the artisans who prepared food (Chap. 7). As social conditions evolved, scientists became more engaged (Chap. 8), and eventually, as the food industry expanded, they were organized to serve its needs (Chap. 9). In this chapter, however, we will see conflict, when scientists outside the industry argue that some of the chemicals used in canning were unsafe. Before diving into the controversies surrounding BPA, it's worth taking a brief look back to the early twentieth century and the first efforts to regulate which chemicals could be added to food.

10.2 Beginning to Regulate Food Additive Safety[1]

At the end of the nineteenth century, the federal government in the United States had a libertarian "buyer beware" approach to food regulation. Manufacturers could, and did, add a dazzling range of undeclared colors and chemical preservatives to their foods to make them more appealing. If these additives were deceptive or even dangerous, then the only remedy for consumers was avoid purchasing the food. The relationship between the consumer and the manufacturer was limited to the marketplace, and government had no role in it.

People began to see this lack of oversight as both unfair and dangerous. After all, if manufacturers weren't required to disclose what they added to food, how could consumers make informed choices? Wasn't it reasonable to expect that something sold as food should not be poisonous? Upton Sinclair's 1906 novel *The Jungle*, which depicted the harsh lives of poor immigrants in Chicago, captured the public mood: "How could they find out that their tea and coffee, their sugar and flour, had been doctored; that their canned peas had been colored with copper salts, and their fruit jams with aniline dyes? And even if they had known it, what good would it have done them, since there was no place within miles of them where any other sort was to be had?" [6]. A coalition of Progressive voices came together as the Pure Food Movement and demanded stronger government regulation. One of their leaders was already inside government: Harvey Washington Wiley, the crusading head of chemistry at the USDA.

Before joining the USDA as chief chemist, Wiley had trained in Germany and taught chemistry at Purdue University. It while he was at Purdue that he began his work on food, using chemical analysis to reveal that honey was often adulterated with added sugars. When he moved to USDA, he expanded this research, and it came to define his career. To Wiley, adding chemicals to food was inherently deceptive: paprika naturally loses its bright red color over time, but artificial dyes could make old spices appear fresh; milk spoils quickly, but a splash of formaldehyde could mask that fact. In his view, food had certain expected natural qualities, and manufacturers were using chemicals to mislead the public about what they were consuming. Food adulteration was wrong because it was, in effect, lying. An important but entirely separate question was whether these chemicals were also dangerous. Wiley, along with many others at the time, believed they were and set out to prove it.

Wiley organized a group of young men from his Washington office to dine together in a special canteen. There, a chef prepared meals which were deliberately spiked with the food additive under investigation. If the participants became ill, that would prove the additive was dangerous. Wiley got the results he both anticipated and wanted. All but one the food preservatives he tested made the participants ill

[1] The history of the federal regulation of chemicals in food is described by Merril [2] and White [3]. The story of Wiley's role in the passage and enforcement of food laws is told by Blum [4] and Rees [5].

(i.e., borax, sulfites, benzoic acid, and formaldehyde. Salicylic acid was the only compound tested that Wiley found less harmful than he expected). Wiley acknowledged that the doses administered were high but argued that such quantities were necessary to observe effects in a reasonable timeframe. After all, he reasoned, children and the infirm would be more vulnerable and subject to chronic exposure to a wider array of chemical additives while his team was exclusively made up of healthy young men.

While Wiley is rightly remembered as a hero in the history of food regulation, there are reasons to question the sense of objective certainty that his scientific experiments conveyed. He was motivated to study the safety of food additives because he already believed they were unsafe. After all, his group was famous at the time as The Poison Squad and the sign that hung in their dining room—"only the brave dare eat the fare"—added to the drama of the situation. From a modern perspective, it's easy to see how his experimental design introduced bias.[2] Squad members knew they were being fed a substance expected to make them ill, and they knew the results their boss wanted. There was no placebo control, which likely encouraged participants to report symptoms regardless of how they felt. Additionally, after participants began avoiding specific foods in the meal they suspected were spiked, the additives were sometimes administered in capsule form during meals. This method could cause localized stomach irritation due to the high concentration released when the capsule ruptured—an effect not representative of consuming the same amount of additive dispersed throughout a food. Many of the substances Wiley deemed harmful, including benzoic acid and sulfites, are now considered safe when used within regulated limits.

The publicity surrounding the Poison Squad's experiments helped fuel the public outcry that led to the passage of the first federal regulation of food additives: the Pure Food and Drugs Act of 1906. The Act defined food as adulterated—and therefore subject to prosecution—if it contained "any added poisonous or other added deleterious ingredient which may render such article injurious to health." It also applied to foods that had been processed or altered in a deceptive manner or that were otherwise "filthy, decomposed, or putrid." However, the Act did not grant the USDA authority to preapprove additives. Instead, the agency could only act against products already on the market. In practice, USDA chemists would analyze a food product and make a recommendation to the Secretary of Agriculture, who could then hold hearings with the manufacturer and potentially initiate prosecution.

[2] Not only from a modern perspective. As Rees points out, other scientists at the time were dismissive of Wiley's approach. For example, the German toxicologist Lehman noted the unrealistically large doses used in Wiley's Poison Squad studies, the scope for psychological bias ("there is no doubt that in Wiley's laboratory preservatives are considered with extreme mistrust and antipathy and that the test persons should imbibe some of this antipathy is easily conceivable"), and limits in the ways Wiley measured changes in physiology ("Careful investigation of the urine, feces and metabolism did not show a single symptom due to the partaking of benzoic acid might be deduced!") [7].

Under the Pure Food and Drugs Act, food additives could be deemed illegal if they were either harmful or deceptive—but the Act did not clearly define what either term meant. Even with Wiley's seemingly objective scientific methods, proving harm was far from straightforward. Wiley himself continued to believe that chemical additives were both deceptive and dangerous—even when he lacked Poison Squad data to support the latter claim. Historian Jonathan Rees quotes Wiley in 1909: "I thoroughly believe in the position I have taken, and I have an abiding conviction that it will be sustained in the end and chemicals of every kind will be excluded from foods" [5].

Other voices, both in industry and government, were willing to see benefit in some additives: synthetic vanilla flavors made ice cream cheaper, benzoic acid prevented spoilage and waste, and caffeine was fine in tea and coffee so it was also fine in Coca-Cola. Wiley objected to all of these—synthetic vanilla was pretending to be something it wasn't, benzoic acid could be harmful in large quantities so should be avoided absolutely, and caffeine was acceptable as a natural part of tea and coffee but unexpected and dangerous outside of that context. Disagreements over the enforcement of the Act eventually led to Wiley's resigning from the USDA in 1912, but he continued to work as a consumer safety advocate, writing for *Good Housekeeping* for the rest of his life.

While much of the food industry initially opposed the new regulations, viewing them as unwarranted government intrusion into the ways they conducted their business, many canners supported the legislation and even saw it as a potential commercial advantage [8]. Trade associations within the canning industry, in particular, recognized that regulation could raise standards and enhance the reputation of the industry as a whole (see Chap. 9). Wiley himself thought canning was one of the few acceptable ways of preserving food as it did not depend on the use of chemicals. Speaking to a meeting of canners in 1906, he described their method "nearest to being the ideal method of preservation" [9].

However, Wiley was not uncritical. He condemned practices such as using short weights or unbalanced ratios of food to brine or syrup. He also expressed concern about the use of artificial colorants intended to make canned foods appear fresher, for example, copper salts in green vegetables and sulfites as bleaching agents in corn.

Another area of conflict between Wiley's USDA and the canning industry involved the long-standing concerns about metals leaching from cans and contaminating food. Lead from solder had been a major concern, but this had largely been addressed with the introduction of Ams cans at the start of the century (see Chap. 4). The other issue was tin.[3] In 1910, Otto Herner, a British public analyst, reported etching of the metal and high levels of tin in British canned fruit. He warned: "I consider the preservation of acid fruits in tin canisters to be entirely improper and fraught with danger to the health of the community unless the canner takes some means to protect the inner surface of the canister, by application of varnish or lacquer, from the attack of the acid" [11].

---

[3]The "salts of tin" issue is described in Chapter 1 of Anna Zeide's dissertation [10].

This report immediately reached the National Canners Association in Washington DC, who worried about the reputational risk to their industry, even though the suspect cans were produced thousands of miles away on the other side of the Atlantic. They sent some American cans to Herner for analysis and were reassured when the results showed much lower levels of tin. Still, the concern remained: any leached tin could be considered an inadvertent food additive and, under the 1906 law, potentially subject to prosecution.

The USDA became interested in the issue, and Willard Bigelow, one of Wiley's senior lieutenants, conducted an extensive survey of tin in American canned foods. His findings showed that while most products contained some tin, the amounts varied widely depending on factors such as the type of tinplate used and the acidity of the food [12]. That was enough to raise concern for Wiley, who planned to issue a public USDA bulletin warning consumers about the presence of tin in canned foods.

The National Canners Association feared bad publicity above all and pushed back against the USDA's proposed warning. While they acknowledged that "salts of tin" could pose a safety risk in certain cases, they argued that such risks were limited to a few isolated instances. Issuing a broad warning about canned foods as a category, they contended, would unnecessarily alarm consumers and damage the industry's reputation. Instead, the NCA proposed collaborating with the government to reduce tin leaching, including promoting the use of protective lacquers and varnishes inside cans where needed. The USDA adopted this approach and rather than issuing the public bulletin instead issued new regulations limiting tin in food to 300 parts per million.[4]

Wiley was generally satisfied with this collaborative relationship, writing later that "the National Canners Association has been foremost amongst the food industries in its endeavor to improve their output by the selection of good material, by sanitary factory methods the improvements of the quality of the container, and the abolition of the use of preservatives" [9]. The cozy relationship between regulators and regulated is further illustrated by the fact that, soon after conducting his survey of tin in canned foods for the government, Willard Bigelow moved to a scientific leadership role at NCA where he worked on the same issue. The salts of tin issue receded from public attention in the 1920s, but similar issues of regulating safety would resurface around BPA from can liners at the end of the century.

10.3 Establishing Additive Safety in Mid-Century

The next major federal law governing food safety was the Food, Drug, and Cosmetic Act, passed in 1938 as part of the New Deal. Like its predecessor, it was a "policing bill," meaning the government could only act against foods already in

[4]The modern limit for tin in canned foods is remarkably similar (250 ppm), and even that is not considered a significant toxicological risk [13]. Higher levels of tin may cause gastric distress and nausea.

commerce—it did not require preapproval of food additives. One important exception was through the introduction of "standards of identity," which allowed the government to define acceptable ingredients for specific food products. This provision was how sodium benzoate came to be effectively banned from tomato ketchup—not because it was proven harmful but simply because it was not included in the official standard of identity for that product.

The 1938 Act also brought greater clarity to the difficult question of what made a food additive "dangerous." Reflecting Wiley's concerns about what was considered "natural," the law treated added chemicals more strictly than toxic substances naturally present in food. Additives were prohibited if they might pose a health risk, whereas naturally occurring toxins were permitted so long as they were not present in amounts that would "ordinarily render [the food] injurious to health." For example, caffeine in coffee was acceptable, but the same compound would face greater scrutiny if added to a soft drink. In both cases, however, the principle remained that "the dose makes the poison"—even a hazardous chemical could be considered safe if used at sufficiently low levels.

The 1938 Act did not specify how food additive safety should be established. However, after the Second World War, scientists from the Food and Drug Administration (FDA[5]) began collaborating with industry researchers to develop standardized testing methods [14]. Their joint effort resulted in the publication of *Procedures for the Appraisal of the Toxicity of Chemicals in Foods* in 1949, a guidance document that has been regularly updated ever since [15]. This work did not attempt to specify which ingredients are safe but rather set down the rules by which safety could be established.

Determining marginal harms was inherently challenging. Effects in humans under realistic dietary conditions were likely to be subtle and slow to appear, while Wiley's earlier approach of administering large doses to observe rapid effects raised other concerns. Instead, the FDA came to favor animal feeding trials. The *Procedures* manual provided detailed recommendations on study design, including the number and type of animals, dosing regimens, study duration, and health outcomes to monitor.

Typically, animals were given a range of relatively high doses, and the lowest dose at which an adverse effect was observed (per unit of body weight) was used to calculate a human safety threshold by applying a 100-fold safety margin. While this was not a direct measure of human safety, the toxicology community considered the margin a reasonable buffer. By first agreeing on the rules, regulators and scientists created a framework for making objective, evidence-based decisions on individual additives.

[5] The implementation of the 1907 Act was initially through Wiley's Bureau of Chemistry at USDA, often in competition with other groups within the agency. The agency was reformed as the Food, Drug, and Insecticide organization in 1927 and renamed the Food and Drug Administration in 1930. In 1950, FDA's responsibilities were moved to the Department of Health and Human Services where they remain.

In the years following the Second World War, growing concerns about the proliferation of synthetic chemicals—especially the pesticide DDT—eventually led to passage of the Food Additives Amendment of 1958. This legislation marked a significant shift by requiring government approval for new food additives following a safety evaluation. However, it included a key exemption: ingredients already in use before 1958, or those recognized as safe by qualified experts, were not subject to the same scrutiny. These exempted substances were classified as generally recognized as safe (GRAS).

If a new ingredient was not considered GRAS, a company would provide toxicological data, established according to standard methods, for evaluation by the government. It wasn't necessary for the additive be complete free from hazard, rather that it did not cause harm at the concentration proposed. One exception to this rule was carcinogenic compounds which the 1958 legislation banned absolutely under the famous Delaney clause: "the Secretary of the Food and Drug Administration shall not approve for use in food any chemical additive found to induce cancer in man, or, after tests, found to induce cancer in animals." Food additives that were toxic at high concentrations were allowable if they were shown to be safe at the concentrations proposed for use; food additives which were carcinogenic under any circumstances were prohibited under all circumstances.

The 1958 law explicitly classified chemicals that leached into food from packaging materials as food additives, providing a regulatory basis for evaluating the epoxy resins increasingly used as can liners. While the resins themselves did not migrate into food, they contained small amounts of unpolymerized BPA, which did. Initially, there appeared to be little cause for concern: BPA was considered only mildly toxic and was rapidly excreted after ingestion, and the quantities involved were small.

However, questions arose about BPA's potential as a carcinogen. When its carcinogenicity was studied in the 1970s, researchers tested adult rodents by feeding them BPA over extended periods at doses just below the threshold for observable toxicity. By those standards, there was no convincing evidence that BPA caused cancer, allowing it to be regulated as an additive rather than banned outright as a carcinogen. The safe exposure level for BPA was calculated by the usual procedure of identifying the lowest dose that caused any adverse effect in animal studies and applying the 100-fold safety margin resulting in an acceptable daily intake of approximately 4 mg for an average adult.

Since typical human exposure was estimated to be more than a thousand times lower than this threshold, BPA was deemed safe. Yet, doubts remained. Perhaps BPA posed risks that the testing methods of the time were not designed to detect.

10.4 A New Class of Risk[6]

In the 1930s, British medical researchers were searching for synthetic estrogen compounds as drug candidates [20]. They injected solutions of various chemicals, including BPA, into five female rats whose ovaries had been removed and found most of them, including BPA, triggered an artificial estrous response. BPA wasn't pursued as a drug, but a closely related compound used in the same study, diethylstilbestrol, was used to prevent miscarriages from the 1940s to the 1960s. However, it was withdrawn in 1971 after it was shown to cause cancer in women treated as well as to harm their children [21]. Diethylstilbestrol was also widely used as a growth promoter in cattle, and this use was not phased out in the United States until 1979, despite concerns about it carrying over to humans in milk and meat. What could this compound be doing to cause harm at low concentrations and across generations?

Cellular processes in plants and animals are regulated by hormones—molecules released in tiny amounts in one part of the body that travel to other tissues to trigger complex biological responses. In mammals, estrogens are a group of hormones that bind to receptors in the ovaries, uterus, and breasts, where they activate specific genes and initiate changes such as the menstrual cycle and the physical transformations of puberty. Estrogens, along with other hormones produced by the mother, fetus, and placenta, also play essential roles in orchestrating key stages of embryonic development.

What was so remarkable about BPA, diethylstilbestrol, and the other compounds in the 1930s study was that these were completely synthetic compounds—many of which had never existed on the planet until a few decades earlier. And yet the rat's physiology recognized them and responded as if they were natural hormones evolved over millions of years. They came to be known as endocrine disrupters because of their effects on the natural hormonal system of the body. Diethylstilbestrol was harmful to women who took it because it triggered hormonal changes that led to cancers. The fact these women's children were also affected showed it could cross the placental barrier and trigger changes in their development too. BPA was a closely related compound, known since the 1930s to be an estrogen mimetic. In the 1970s, BPA had been shown to diffuse into the food at low levels, but, because it was neither toxic nor carcinogenic according to the standard methods used, it wasn't seen as an important problem. It wasn't until the 1990s that people began to worry its hormonal activity in food could be a new class of risk.

A research team at Stanford University wanted to know if baker's yeast produced estrogenic compounds [22]. They grew their yeast in a nutrient broth prepared with

[6] Sarah Vogel's doctoral dissertation [16] on the history of BPA and the disputes over its safety along with her related papers [17, 18] form the basis for much of this section, along with Chapter 10 of Anna Zeide's book [10]. The long piece written by Trevor Butterworth for the nonprofit STATS.org in 2009 provides a useful counterpoint, one skeptical of the view that BPA is harmful [19].

water which had been previously sterilized in an autoclave (a pressure cooker similar to a canner's retort). As they had hoped, the samples did have estrogenic activity, but there was something strange in their data. First, when more yeast was grown in the same containers, it did not create any more estrogenic activity—surely if the yeast was making the compound responsible, then more yeast would mean more compound? Second, when water was autoclaved without any yeast present, it was equally estrogenic. The yeast wasn't responsible for the estrogenic effect; something in the water was.

The water had been autoclaved in polycarbonate flasks, and when the Stanford researchers looked carefully, they found tiny quantities of BPA had leached into it from the water. This obviously went far beyond a methodological issue for the scientists; it had the potential to be a huge public health concern. The concerns were picked up, not by food toxicologists but by a community of campaigning environmental scientists.

Fred vom Saal was a professor at the University of Missouri. Since his doctoral work at Rutgers, he had been interested in the effects of natural hormones on animal development. He learned that even tiny changes in the hormonal environment of the embryo could change the bodies and behaviors of the adult animal. In one striking example, he showed differences in mice if their "neighbors" in the uterus during gestation were sisters or brothers. Three important insights from this work would come to inform his views on the BPA issue. First, the amounts of hormone needed were tiny, parts per billion or even trillion, and far below what would be considered for toxicological studies. Second, the effect of the hormone did not increase steadily with concentration but rather reached a peak and then declined. Finally, what mattered more than the dose size was when the exposure to the hormone was received. Tiny amounts crossing the placental barrier at a critical phase of embryo development could be seen years later as pathologies in the adult.

Vom Saal came across the issue of synthetic endocrine disruptors in July 1991 at a small, interdisciplinary meeting of wildlife biologists and other environmental scientists concerned with pollution, especially products of the plastics and petrochemical industries [23]. The group concluded their meeting by issuing a consensus statement that "compounds introduced into the environment by human activity are capable of disrupting the endocrine system of animals, including fish, wildlife, and humans" [24]. The Wingspread meeting—named for the conference center in Racine, Wisconsin, where it was held—helped build a community of scientists and activists around the issue of endocrine disruptors. It is this community that proved critical in challenging the scientific consensus that BPA was safe.

The compounds discussed at the Wingspread meeting included DDT, PCBs, dioxins, and diethylstilbestrol (DES)—the compound closely related to BPA that had been disastrously used as a drug—but not BPA itself. However, vom Saal had been inspired by what he learned at Racine and went back to his lab in Missouri determined to study the effects of other synthetic hormones.

His team showed that BPA did indeed bind to human estrogen receptors and did so more strongly than natural estrogen [25]. Even more concerning were the results of animal studies. When pregnant mice were fed tiny, supposedly "safe" doses of

BPA, their adult male offspring developed enlarged prostate glands. These exposure levels were comparable to what human mothers might receive from drinking water stored in polycarbonate bottles or eating food from cans lined with epoxy resins. Other laboratories reported similar findings: young female rats fed BPA at levels within the human safety limit had a faster maturation of their mammary tissues [26], mice whose mothers consumed similar doses of BPA during pregnancy had lower survival rates at birth, and those that survived reached puberty more quickly [27].

Surely this growing body of evidence would compel government action.

10.5 A Scientific Controversy

However, instead of acting, the government first sought more reliable data. In 1996, Congress directed the Environmental Protection Agency (EPA) to investigate the risks of endocrine disruptors, including BPA. However, demonstrating the harm—or establishing the safety—of endocrine disrupters was a challenge to traditional methods.

Before addressing these special difficulties, it's worth returning to why this apparently simple question is so hard to answer. At the core of the issue lies a difficult truth: the definitive experiment needed to answer the question simply cannot be conducted. To understand why, imagine what it might look like: we would take a representative group of people and divide them into two groups. One group, the control, would get a diet containing no BPA, a challenge in itself given how widely distributed the chemical is in the environment; the other group would be given an identical diet with a known amount of added BPA. The experiment is unlikely to meet ethical standards, as it involves intentionally exposing people to a substance believed to be harmful.

If we went ahead anyway, the experiment, with its strictly controlled diet, would have to continue for decades, as the risks may be slow to reveal themselves. As we are particularly concerned with reproductive risk, we might also want to follow the children of the study group members too, perhaps until puberty.

After a time, however, we might begin to see significantly higher disease rates in the group exposed to BPA, and this would be proof of harm. Even if we did not, however, that would not be enough to declare the ingredient safe. Perhaps we should have studied different types of people? It may be women are more vulnerable to BPA than men or overweight people were more vulnerable than the lean. Perhaps we should have considered higher or lower doses of BPA? Any one of these questions would demand another one of these impossible experiments. The only way to stop would be to decide that what you had was "good enough." Although that question involves scientific judgment, it is ultimately a political issue about weighing risks and benefits—to the extent they are known. The situation only gets more complicated as we must reach these decisions in the absence of a human trial.

If the gold standard clinical trial is ethically, logistically, and economically impossible, practical toxicological risk assessments must depend on proxy measurements. A good proxy experiment is something that can be done relatively easily yet accurately predicts the unknown risk of real human exposure. Wiley had tried giving his Poison Squad large doses of additives as a proxy for long-term exposure to lower levels of the same additives, but, as we have seen, this approach had problems.

By mid-century, animal feeding studies were the norm as proxies for toxicological experiments in humans, but these introduced new issues. For example, can we be confident the effect in an animal is the same as the effect in a human? Studies in the 1970s showed the artificial sweetener saccharin caused bladder cancer in rats which, under the Delaney Clause, should have led to its ban as a food additive [28]. However, later research revealed that rats and humans metabolize saccharin differently, undermining the original conclusion. As a result, saccharin was eventually reclassified and returned to the market as an apparently safe sweetener. Toxicologists have responded to these concerns by requiring feeding studies in multiple different species of animal. However, this opens the question of how many—if you see no effect in rats and mice, should you go on to study pigs, monkeys, or chimpanzees? Once again, the only way to stop would be decide that what you had was "good enough."

These are some of the foundational problems in regulator toxicology for any compound. Assessing the risks of endocrine disrupters faced all these and several more besides. First, the sheer ubiquity of potential phytoestrogens in the environment made it difficult to conduct reproducible experiments. Were natural components in soy-based rodent feed influencing the results? And to what extent was laboratory water already contaminated with synthetic estrogens?

More profoundly, toxicologists had long depended on the idea that "the dose makes the poison," implying that below a certain level, even a known toxin can be taken as harmless. This principle is inherent in the regulation of an "acceptable level" of tin in canned food in the 1910s and in the 100-fold margin of safety used to set the limits of BPA in canned food. However, vom Saal and the endocrinologists argued that sometimes the effect of endocrine disrupters was non-monotonic, meaning a low dose could be more impactful than a higher dose. If true, this would invalidate the toxicologists' method of using high concentrations to observe noticeable effects and then estimating a safe dose as a small fraction of that value.

However daunting the technical questions, the social issues were worse. The reason FDA had been able to reach decisions up to this point had been due, in part, to the communities of expert toxicologists who had agreed on the best methods to determine harm, for example, 1949's *Procedures* manual [15]. Prior agreement on methods allowed apparently objective scientific assessments on a given case: we know this additive is "safe" because these mice did not get sick under these conditions. However, the apparent objectivity of the decision hides the choices that had been made to establish what could only be a highly indirect assessment of safety. Toxicology, as a field relevant to regulation, struggles to maintain its authority as objective science distinguishing unambiguously between the absolute categories of "safe" and "unsafe" while at the same time dealing with the realities of political and legal controversy [29].

These are genuine concerns within toxicology, but those most worried about endocrine disruptors were not traditional toxicologists. Instead, they were activist environmental scientists, representing a different community. Communities of scientists are formed over generations through shared educational programs, working together in the same labs, publishing in the same journals, and attending the same meetings. Scientific communities are often social, with lifelong friendships and fierce rivalries established in these venues. They provide a shared sense of "how things should be done" which helps them reach agreement on technical and ethical issues. Without a shared community, agreeing on protocols, let alone reaching consensus and making recommendations, was always going to be hard.

What followed was a predictable "paper war," as scientists from the different communities tried to settle the question with dueling publications in the peer reviewed literature. Workers in vom Saal's laboratory had shown that low doses of BPA fed to pregnant mice led to larger prostates in their male offspring [25]. When other teams of scientists from industry tried to replicate this finding, they saw no such effects [30, 31]. Endocrinologists, including vom Saal, found fault with the replication study [32]. Regulatory toxicologists in turn rebutted their rebuttal [33]. The points of dispute became so arcane—who is truly qualified to dissect out the prostate gland of a mouse?—that few outsiders could hope to understand, let alone form a scientific opinion. One gets the sense that the endocrinologists saw the toxicologists as hiding behind outdated methods to support their industry paymasters, while the toxicologists saw the endocrinologists as playing fast and loose with methods to justify a decision they had already reached, affecting an industry they did not like. For policymakers, none of this was helpful; science was not reaching a consensus that might guide regulation.

To try to resolve issue, expert panels of scientists were formed by the industry, by environmental campaigners, and by government bodies to evaluate the available data on BPA and reach a consensus recommendation. However, the process quickly became contentious. First of all, who even counted as an expert? Scientists with extensive contacts either with industry or the environmental movement could easily be seen as biased and unable to give an objective judgment. Yet the specialized nature of the topic meant it was difficult to find qualified individuals without such affiliations. Second, which studies should these panels consider as reliable evidence? On this issue, the question of "good laboratory practice" (GLP) was a particular sticking point [32].

GLPs are part of a system of recordkeeping and quality control checks designed to ensure the integrity and reliability of measurements used in regulatory decision-making.[7] However, GLP standards are expensive and cumbersome to maintain and,

[7] Ironically, the 1970s scandal that led to GLPs being required in the first place involved the contract laboratories whose work had shown that BPA was not carcinogenic. They were accused of using fraudulent and shoddy research methods and even simply fabricating data. In response to these concerns, the FDA introduced regulations in the early 1980s requiring the use of good laboratory practices (GLPs) as a quality control measure for scientific work that would be used in regulatory decision-making.

in practice, are typically followed only in specialized commercial laboratories whose results are intended for use in legal or regulatory contexts. University laboratories, even when producing high-quality research, generally do not adhere to GLPs. This creates a divide: industry-funded studies are often conducted in commercial labs under GLP conditions, while federally funded research is usually carried out in universities without GLP compliance. Critics of GLPs argued that insisting on their use effectively excluded much of the work done by people concerned about endocrine disruptors and noted that the funding source seemed to reliably predict the results. In one review, 94 of the 115 government-funded studies showed low doses of BPA caused an effect, but none of the 11 industry-funded studies did [34]. In response, proponents of GLPs acknowledged that while the smaller-scale, more innovative approaches used in university laboratories are valuable for identifying potential hazards, regulatory decisions require rigorous standards and validated methods. GLPs, they argued, provide the necessary framework to ensure consistency, reliability, and accountability in scientific research used for policy-making [33]. In their view, one or two large and complete studies are more valuable than hundreds of small studies.

Unsurprisingly then, a 2000 report by the National Toxicology Program (NTP) did include vom Saal's work in their considerations and found there was credible evidence of effects from BPA at low levels. In contrast, a 2004 report sponsored by the American Plastics Council did not include vom Saal's work and found low levels of BPA were safe.

Surely though, the controversy could be resolved by just getting the data right? In 2012, the federal government funded a major new study, CLARITY-BPA, to do just that [35]. The study was designed collaboratively to bridge the expectations of different research communities by combining the reliability of GLP standards with the flexibility needed to include university-based researchers. This approach aimed to ensure both methodological rigor and broad scientific participation. Rats were fed controlled doses of BPA in a GLP certified lab, which also did some measurements of the effects (the "core"). Additionally, university researchers were able to participate by receiving blinded samples from the test animals and their offspring. These samples could then be analyzed in non-GLP laboratories using the broader range of methods favored by endocrinologists [23]. The core group reported their findings in 2018, and the final report, including most of the university researchers' results, was released in 2021 [35, 36].

In the core group study, a few measurements did show significant pathologies due to BPA exposure but these "were not dose responsive … and did not demonstrate a clear pattern of consistent responses within or across organs." An analysis of the core study data by FDA scientists concluded that, "on a weight-of-evidence approach, we conclude that the core study data do not suggest a plausible hazard of BPA exposure at the lower end of the dose range tested" [37]. BPA was safe.

Or perhaps not. Several, although by no means all, of the university researchers reported significant effects of low-dose BPA. For example, vom Saal's group reported significant effects of parental BPA exposure on the fetal development of the rats' urinary and reproductive systems [38]. A review of the data by a team of university researchers found "Collectively, the CLARITY results highlighted herein contribute to and strongly corroborate a significant body of evidence that documents adverse effects of BPA at low doses" [39].

Despite all of this, and over the vigorous objections of the endocrinology community [23, 40], BPA remained legal in the United States under FDA regulations [41]. Interestingly though, other jurisdictions reached different conclusions based on substantially the same data. Individual US states have passed restrictions [42]. The EU banned BPA in packaging containers for babies and children in 2018 and, in 2023, issued a scientific opinion that drastically lowered the daily tolerable daily intake of BPA from 4000 ng/kg of body weight to only 0.2 ng/kg [43]. In 2008, the Canadian government took a precautionary approach by adding BPA to a schedule of toxic substances while acknowledging uncertainty in the data. By 2014, Canadian government action had effectively eliminated BPA from packaging materials designed for infants and from infant formula [44]. Clearly the question was being resolved by different political realities rather than different scientific facts, and it is worth reexamining the case through that lens.

10.6 A Political and a Commercial Controversy

It is worth stating the obvious. The plastics and petrochemical companies who made BPA along with their trade associations had a financial interest in the question of whether it was safe. They faced loss of markets if the product was restricted and could also be legally liable if their products were shown to have harmed people. Furthermore, because the criticism of BPA was part of a broader campaign against alleged environmental endocrine disruptors, BPA manufacturers found common cause with companies that produced DDT, dioxins, pesticides, and other chemicals of concern.

There was also, undoubtedly, a psychological dimension to the motivations of scientists working at these companies that extended beyond their professional responsibilities. None of them had entered their careers intending to cause harm and for them to acknowledge that their work may have done so would carry a significant emotional burden. Many sincerely believed in the benefits their products provided and feared the consequences of regulatory bans. For instance, epoxy coatings on cans were genuinely superior, and failures in these coatings really could lead to food spoilage and safety concerns. These sentiments often extended beyond individuals to broader professional communities; a chemist, for example, might perceive that "chemistry" itself was under attack and dismiss concerns about BPA without fully engaging with the evidence. In some cases, these values aligned with a right-libertarian political philosophy that prioritizes individual choice over government intervention.

The companies and individuals who used BPA in their products—here, we are particularly interested in the canning industry—shared many of the same interests as the BPA manufacturers, but not all of them, and not to the same degree. We will return to their distinct perspectives at the end of this section.

It is equally clear that the community of scientists opposing BPA had a vested interest in the outcome of the debate. Although the organizations supporting this viewpoint had less obvious financial incentives, their opponents in industry often suspected ulterior motives such as fundraising for nonprofits or lucrative lawsuits against rich companies. The scientists involved in opposing BPA likely believed their work was protecting people and the environment from the overlooked harms of technology and corporate greed. Indeed, many of them dedicated their careers to this mission. As a result, they may have been dismissive of the potential benefits of the technologies they criticized. These convictions often extended into the broader environmental movement, where individuals concerned with other ecological issues might oppose BPA without closely examining the specific evidence. In many cases, these values aligned with a liberal (in the American sense) political philosophy that emphasizes protecting individuals from corporate power through state intervention.

Both sides engaged in political maneuvering to advance their goals, and the scientific work itself was often explicitly intended to influence policy. The 1991 Wingspread meeting, for example, was organized by Theo Colborn and Pete Myers as part of a broader campaign to raise awareness about endocrine-disrupting chemicals. Colborn, who earned her doctorate in zoology at nearly 60, had worked in both federal agencies and environmental organizations, particularly on pollution issues in the Great Lakes. She later joined the W. Alton Jones Foundation under Myers, where she continued her advocacy. The Wingspread Conference was a continuation of this work, aiming to spotlight the hormonal effects of environmental pollutants. Myers had previously written on climate change and briefly experimented with the term "climate disruption" to add political urgency. When he applied a similar rhetorical strategy and called the compounds he was concerned about "endocrine disrupters," it gained more traction with the public. In contrast, when chemical companies formed a group to present their perspective, they chose a more neutral name: the Endocrine Issues Coalition. Both names were deliberate rhetorical choices.

The Wingspread consensus statement enabled the participating scientists to affirm the areas where they had strong agreement, rather than focusing on their points of disagreement. This was politically useful not only in uniting the participants into a cohesive movement but also by signaling to lawmakers that certain aspects of the issue were "settled science," so no further research was needed before acting. Coburn immediately put the consensus statement to work in Washington, meeting with politicians and even carefully timing a flight connection so she could travel with the Administrator of the FDA and share a copy with him. Wingspread participants testified at congressional hearings about the effects of estrogens. The BBC made a documentary (*The Estrogenic Effect: Assault on the Male*) which went on to win an Emmy Award—Colburn and Myers collaborated on a popular science book (*Our Stolen Future*—the Wingspread consensus statement was included), and Vice President Al Gore wrote the foreword.

The Wingspread Conference also helped form a generation of scientist-activists. Before the meeting, vom Saal was a respected research scientist rather than a campaigner, but he grew increasingly frustrated with the lack of action on what he saw as clear evidence of harm. He turned to the media to make his case. His work was featured in *USA Today* in November 2007, in April of the following year in *Discover* magazine, in May on PBS, and in June in *Time*. Vom Saal was featured in the series of stories on the plastics issue in *The Milwaukee Journal Sentinel* which went on to win a George Polk Award. In 2009, he was interviewed for a documentary film on the issue—*Plastic Planet.*

Much of the media coverage was sympathetic to the environmentalists' concerns and supported a view that BPA was a case of industry deliberately trying to hide the harms from their products. Very little attention was paid to the points about the importance of GLPs and reproducibility the chemical industry kept trying to raise. Perhaps this was a sophisticated read of the data by journalists, or perhaps their political viewpoints coincided with those of the environmentalists. As the journalist Trevor Butterworth argued at the time: "[b]ecause the charge of industry bias resonates so powerfully with journalists, it often serves to obscure both the fiscal realities and methodological rigor of scientific research. In other words, it frees journalists from the need to demonstrate which studies are correct and which are not" [19].

This was all politics. Colburn and Myers had seen the environment harmed by pollution from the wealthy and influential chemical industry and recognized that effective opposition needed more than plain facts. They had organized the Wingspread Conference, because science is valued as a source of truth in society and the consensus of a group of scientists carries yet more authority. They had first taken their concerns directly to legislators. Now they were fighting to shape public opinion, with the hope that would influence the decisions of elected policy-makers. Critics would say they were "politicizing science."

At the same time, chemical manufacturers sought to defend their products by casting doubt on the emerging science. Much of this effort was coordinated indirectly through trade associations and through the apparently objective public advocacy groups they quietly funded. They argued that the classification of their products as endocrine disrupters was merely a hypothesis and that further research—and thus delay—was necessary before implementing any drastic regulatory measures. They also deflected attention by pointing to the presence of many naturally occurring estrogens in the environment. The book, *Our Stolen Future*, was dismissed by critics as "junk science."

Their efforts put elected officials responsible for regulation in a difficult position; whatever decision they took on BPA would upset a valuable constituency. The chemical industry, along with its donors, employers, and conservative-libertarian voters, would likely be angered by a ban on BPA—while environmental advocates and liberal voters would be equally upset if it were not banned. Politicians who want to be reelected try to avoid taking unpopular positions and would often prefer to pass the responsibility to experts. Critics would say they were trying to "scientize politics," that is, they were trying to transform a difficult political decision into an apparently objective scientific decision [45].

This approach had worked well for ingredient safety for many years because of the consensus on methods developed by the toxicologists. It broke down when the endocrinologists challenged that consensus. The repeated attempts to form expert panels, for example, the CLARITY-BPA program, were efforts to form a new scientific consensus that could take the BPA decision out of the political realm. What they achieved, however, was more firmly entrenching politics in the scientific realm. While the meetings would be on apparently scientific questions like GLPs, both sides knew the political implications and proceeded accordingly. For example, Vogel describes Theo Colburn's recollection that industry scientists would consult with industry lawyers during breaks in a what was claimed to be a scientific meeting [16].

Ultimately, though, these sorts of decisions must be made politically. Science tries to determine how the world is—the risks of BPA—but it is our values, expressed politically, that determine what we ought to do in response. After all, about 40,000 Americans die on the roads each year and ten times that number from the results of smoking, yet cars and tobacco both remain legal. The harms from BPA may be profound but are far less clearly understood. Clearly the automobile, tobacco, and petrochemical industries all try to influence political decision-making, but that does not diminish the fact that the decision is political. How you feel about that corporate influence, indeed how you feel about the issue of BPA today, is more likely to be determined by your political viewpoint than on any detailed understanding of the science involved.

If the story of BPA was a conflict between the chemical industry and environmentalists, it is worth noting that the canners always sat slightly to one side of the main fight. The canners used BPA in their linings and so shared some of the commercial and individual interests with the people who made it. However, there were some important differences. The companies who made BPA had to defend their product, because that was the only way they could defend their business. Their resins were certainly useful to the canners, but they were not the only option. Although the canners might sincerely believe that BPA was safe, they recognized that consumers' opinions were even more important. As one trade journal noted in 2011, if "consumers perceive there is a potential hazard, … that is a significant issue" [46]. If BPA in cans became a reason for consumers to stop buying canned foods, then epoxy liners were no longer useful. Food historian Anna Zeide illustrates this dual position by describing how, during this period, Campbell's Soup would simultaneously reassure consumers that it was working to remove BPA from its cans while also supporting its trade association's efforts to allow the chemical's continued use [10].

Although the campaign to sway public opinion against BPA did not gain enough momentum to override commercial interests in the legislative process, it did succeed in making consumers think twice about buying canned food. That was enough to make the canners quickly demand alternatives from the coatings manufacturers. Epoxy resins produced using compounds other than BPA were marketed as can linings from the early 2000s. Many of these polymers were formed with closely related compounds (e.g., bisphenol F and bisphenol S), but others were polymers formed with different starting materials (e.g., diphenyl sulfone). Canning without BPA became a marketing advantage that companies would exploit in advertising to differentiate their products from their competitors.[8]

By 2019, although BPA remained legal, it had been phased out of can liners in the United States and many other countries [47]. Whether this will result in a public health benefit remains uncertain, as the replacement compounds are far less studied than BPA and several have been found to be estrogenic or harmful at very low concentrations [48–50].

10.7 Reflections for Modern Scientists: Science and Politics

In this chapter and in the last, we have seen several examples of ways the canning industry has responded to challenges over the safety of its products. When the industry's reputation was threatened by highly publicized botulism cases, it launched a major scientific initiative to determine with certainty what constituted adequate cooking (see Chap. 9). Although the results were more ambiguous than publicly acknowledged, the effort nonetheless contributed to improved processes and helped shape a broader system of regulations. When concerns arose about "salts of tin" around the same time, the industry was wary that Wiley would issue a public warning that might alarm consumers about canned food in general. However, they were willing to cooperate quietly to address the issue.

In both the tin and botulism cases, the canners, through their trade associations, were eager to work with regulators—even if it meant opposing fellow canners whose lower standards threatened the industry's reputation. This collaborative approach proved effective, not because it guaranteed that all cans would forever be free of botulism or tin but because the involved parties agreed that the steps taken were reasonable. As a result, both issues faded from public concern, and the canners breathed a collective sigh of relief. The BPA case, however, unfolded differently and the way it did suggests alternative roles for scientists today.

First, while consensus around methods is important for regulatory decision-making, independent groups of scientists play a crucial role in questioning the foundations of that consensus. The toxicology community, relying on animal studies and using what they considered an abundance of caution, had concluded that BPA was safe; it was "settled science." In contrast, environmental scientists employed different methods and reached different conclusions. Both approaches are inherently indirect in assessing human harm, yet each community believed its own methodology was both scientifically valid and the appropriate basis for regulation. Conflict was inevitable, and the outcomes were messy. However, without external scientific

[8] Initially many of these were marketed as "BPA-free," but this has moved toward claims such as "made without BPA." BPA has become so widespread in the environment that is inevitably present at some level in can liners (and everything else) even if it was not intentionally added.

voices willing to challenge the prevailing orthodoxy, there would be no opportunity to "unsettle the science" and drive reform.

Second, outside groups play a vital role in bringing attention to issues that might otherwise go unnoticed. Some problems are highly visible and demand immediate response. For example, the botulism cases from commercially canned foods in the 1910s were tragic, but relatively rare compared to other risks of the time. Still, when they occurred, they were sensational enough to generate public alarm and prompt swift action from both industry and government. Other problems, however, are nearly invisible and require sustained advocacy to build the political will for change.

If the health effects of BPA from canned food are as serious as some believe, the issue represents a far greater public health threat than botulism ever did. Yet BPA's victims are not identifiable as individuals. If we say BPA causes birth defects, we mean that there is a statistically higher incidence among children born to mothers exposed to it. We cannot claim that any individual child's condition can be definitively traced to the chemical. Many other factors, including other environmental chemicals, also contribute to risk. Because the harm is distributed statistically across the population, it is harder to recognize and mobilize public opinion. Similar challenges have arisen in public health campaigns against smoking or lead exposure. In such cases, scientists can serve a crucial social function by using their data to raise awareness and generate public pressure for reform.

Finally, it is worth reflecting on how industry responds to these challenges as both threats and opportunities. When the threat affects the entire industry rather than a single company, trade associations often take the lead. They typically engage in the scientific debate while simultaneously leveraging that science in public relations campaigns aimed at shaping consumer opinion. At the same time, individual companies may view the situation not as a threat, but as a market opportunity. When an additive or process becomes controversial, there is often a chance to voluntarily introduce an alternative that addresses public concerns. For instance, in the early twentieth century, companies that adopted "sanitary cans" made by Ams machines could market their products as safer, since they avoided exposing consumers to lead from solder. A century later, companies used "BPA-free" can liners to make similar claims, positioning their products as safer by avoiding the perceived risks of endocrine disruptors. In such cases, consumer perception of risk often outweighs expert opinion in shaping market behavior.

References

1. Gannon, J.A.: History and development of epoxy resins. In: Seymour, R.B., Krishenbaum, G.S. (eds.) High Performance Polymers: Their Origin and Development, pp. 299–307. Elsevier Science Publishing Co. (1986). https://doi.org/10.1007/978-94-011-7073-4
2. Merrill, R.A.: Food safety regulation: reforming the Delaney Clause. Annu. Rev. Public Health. **18**, 313–340 (1997)
3. White, S.R.: Chemistry and Controversy: Regulating the Use of Chemicals in Foods, 1883–1959. Emory University (1994)

4. Blum, D.: The Poison Squad. Penguin Press (2018)
5. Rees, J.: The Chemistry of Fear: Harvey Wiley's Fight for Pure Food. Johns Hopkins University Press (2021)
6. Sinclair, U.: The Jungle. Doubleday, Page & Company (1906)
7. Lehman, K.B.: Most recent investigations on the determination, preservative action and admissibility of the use of benzoic acid. Science. **35**, 577–585 (1912)
8. Pearson, G.S.: The Democratization of Food: Tin Cans and the Growth of the American Food Processing Industry, 1810–1940. Lehigh University (2016)
9. May, E.C.: The Canning Clan. Macmillan Publishers Limited (1937)
10. Zeide, A.: In Cans We Trust: Food, Consumers, and Scientific Expertise in Twentieth-Century America. University of Wisconsin-Madison (2014)
11. Anonymous.: The dangers of canned fruit. Lancet. **175**, 1390 (1910)
12. Bigelow, W.D., Bacon, R.F.: Bureau of Chemistry – Circular No. 79. Tin Salts in Canned Foods of Low Acid Content: With Special Reference to Canned Shrimp. U.S. Government Printing Office (1911)
13. Blunden, S., Wallace, T.: Tin in canned food: a review and understanding of occurrence and effect. Food Chem. Toxicol. **41**, 1651–1662 (2003)
14. Stirling, D., Junod, S.: Profiles in toxicology: Arnold J. Lehman. Toxicol. Sci. **70**, 159–160 (2002)
15. Lehman, A.J., Laug, E.P., Woodard, G., Draize, J.H.: Procedures for the appraisal of the toxicity of chemicals in foods. Food Drug Cosmet. Law Q. **4**, 412–434 (1949)
16. Vogel, S.A.: The Politics of Plastics: The Economic, Political and Scientific History of Bisphenol A. Columbia University (2008)
17. Vogel, S.A.: The politics of plastics: the making and unmaking of bisphenol A 'safety'. Am. J. Public Health. **99** (Suppl 3), 559–566 (2009)
18. Vogel, S.A.: From 'the dose makes the poison' to 'the timing makes the poison': conceptualizing risk in the synthetic age. Environ. Hist. **13**, 667–673 (2008)
19. Butterworth, T.: Science suppressed: how America became obsessed with BPA, pp. 1–45 (2009). Available at: https://web.archive.org/web/20260211130416/https://h2o-online.co.za/wp-content/uploads/2020/07/STATS_-_Science_Suppressed_-_Americas_Obsession_with_BPA_-_June_12_20091.pdf
20. Dodds, E.C., Lawson, W.: Synthetic estrogenic agents without the phenanthrene nucleus. Nature. **137**, 996 (1936)
21. Veurink, M., Koster, M., De Jong-Van Den Berg, L.T.W.: The history of DES, lessons to be learned. Pharm. World Sci. **27**, 139–143 (2005)
22. Krishnan, A.V., Stathis, P., Permuth, S.F., Tokes, L., Feldman, D.: Bisphenol-a: an estrogenic substance is released from polycarbonate flasks during autoclaving. Endocrinology. **132**, 2279–2286 (1993)
23. Soto, A.M., Schaeberle, C.M., Sonnenschein, C.: From wingspread to CLARITY: a personal trajectory. Nat. Rev. Endocrinol. **17**, 247–256 (2021)
24. Statement from the work session on chemically-induced alterations in sexual development: the wildlife/human connection. Adv. Mod. Environ. Toxicol. **21**, 1–8 (1992). https://pmc.ncbi.nlm.nih.gov/articles/PMC1469664/
25. Nagel, S.C.: Relative binding affinity-serum modified access (RBA-SMA) assay predicts the relative in vivo bioactivity of the xenoestrogens bisphenol A and octylphenol. Environ. Health Perspect. **105**, 70–76 (1997)
26. Colerangle, J.B., Deodutta, R.: Profound effects of the weak environmental estrogen-like chemical bisphenol A on the growth of the mammary gland of noble rats. J. Steroid Biochem. Mol. Biol. **60**, 153–160 (1997)
27. Howdeshell, K., vom Saal, F.S.: Developmental exposure to bisphenol A: interaction with endogenous estradiol during pregnancy in mice. Am. Zool. **40**, 429–437 (2000)
28. Price, J.M.: Bladder tumors in rats fed cyclohexylamine or high doses of a mixture of cyclamate and saccharin. Science. **167**, 1131–1132 (1970)

29. Demortain, D.: Regulatory toxicology in controversy. Sci. Technol. Hum. Values. **38**, 727–748 (2013)
30. Cagen, S.Z.: Normal reproductive organ development in CF-1 mice following prenatal exposure to bisphenol A. Toxicol. Sci. **50**, 36–44 (1999)
31. Ashby, J., Tinwell, H., Haseman, J.: Lack of effects for low dose levels of bisphenol A and diethylstilbestrol on the prostate gland of CF1 mice exposed in utero. Regul. Toxicol. Pharmacol. **30**, 156–166 (1999)
32. Myers, J.P.: Why public health agencies cannot depend on good laboratory practices as a criterion for selecting data: the case of bisphenol A. Environ. Health Perspect. **117**, 309–315 (2009)
33. Tyl, R.W.: Basic exploratory research versus guideline-compliant studies used for hazard evaluation and risk assessment: bisphenol A as a case study. Environ. Health Perspect. **117**, 1644–1651 (2009)
34. vom Saal, F.S., Hughes, C.: An extensive new literature concerning low-dose effects of bisphenol A shows the need for a new risk assessment. Environ. Health Perspect. **113**, 926–933 (2005)
35. Schug, T.T.: A new approach to synergize academic and guideline-compliant research: the CLARITY-BPA research program. Reprod. Toxicol. **40**, 35–40 (2013)
36. NTP: NTP research report on the ClARITY-BPA core study: a perinatal and chronic extended--dose-range study of bisphenol A in rats. NTP Res. Rep. Ser. **9**, 1–121 (2018)
37. Camacho, L.: A two-year toxicology study of bisphenol A (BPA) in Sprague-Dawley rats: CLARITY-BPA core study results. Food Chem. Toxicol. **132**, 110728 (2019)
38. Uchtmann, K.S.: Fetal bisphenol A and ethinylestradiol exposure alters male rat urogenital tract morphology at birth: confirmation of prior low-dose findings in CLARITY-BPA. Reprod. Toxicol. **91**, 131–141 (2020)
39. Prins, G.S., Patisaul, H.B., Belcher, S.M., Vandenberg, L.N.: CLARITY-BPA academic laboratory studies identify consistent low-dose bisphenol A effects on multiple organ systems. Basic Clin. Pharmacol. Toxicol. **125**, 14–31 (2019)
40. Vom Saal, F.S., Vandenberg, L.N.: Update on the health effects of bisphenol A: overwhelming evidence of harm. Endocrinology. **162**, 1–25 (2021)
41. FDA: Bisphenol A (BPA). Food Additives and Petitions (2023) Available at: https://web.archive.org/web/20230403153623
42. Kadasala, N.R., Narayanan, B., Liu, Y.: International trade regulations on BPA: global health and economic implications. Asian Dev. Policy Rev. **4**, 134–142 (2016)
43. EFSA. Bisphenol A (2023). Available at: https://web.archive.org/web/20230727191404/
44. Health Canada. Bisphenol A (BPA) risk management approach: performance evaluation for BPA-HEALTH component (2018). Available at: https://web.archive.org/web/20230727192215/
45. Pielke, R.A.: The Honest Broker: Making Sense of Science in Policy and Politics. Cambridge University Press (2007)
46. Brody, A.L.: Taking a closer look at BPA and its alternatives. Food Technol. **65**, 79–81 (2011)
47. Can Manufacturers Institute: CMI Washington State Canned Food Market Basket Survey. CMI (2020)
48. Harnett, K.G., Chin, A., Schuh, S.M.: BPA and BPA alternatives BPS, BPAF, and TMBPF, induce cytotoxicity and apoptosis in rat and human stem cells. Ecotoxicol. Environ. Saf. **216**, 112210 (2021)
49. Fouyet, S., Olivier, E., Leproux, P., Dutot, M., Rat, P.: Bisphenol A, bisphenol F, and bisphenol S: The bad and the ugly. Where is the good? Life. **11**, 1–13 (2021)
50. Kodila, A., Franko, N., M.S., D.: A review on immunomodulatory effects of BPA analogues. Arch. Toxicol. **97**, 1831–1846 (2023)

Chapter 11
The Continuing Invention of Canned Food

Abstract What we can gain by recognizing the technologies that support us. Virtues of technology workers.

11.1 Really Seeing Tin Cans

Humans are distinguished from other species by our twin dependencies on the technology we make and on one another. The biggest changes in human society are associated with changes in the ways we use technology.

It is perhaps because our day-to-day lives are so deeply shaped by technologies that we are frequently barely aware of their presence. People only notice the sewer line to their house when it starts to smell. Instead, we focus on new technologies that may disrupt our lives tomorrow. Inventors and entrepreneurs are heroes to some, while others protest against the harms their innovations bring.

This book followed the story of a single technology—canning—not only its invention but through the centuries of iterative change that followed. We saw canning transform from something new and startling in Paris in 1810 to everyday and invisible in Andy Warhol's 1960s New York. As canning technology evolved, canned food became progressively more important in peoples' diets and, at the same time, less recognized as a technology by the general public.

This combination of increased economic importance and decreased social visibility generally suited the interests of the canning industry. When the technology was questioned and became "visible"—such as during the botulism or BPA crises—the industry responded swiftly. Still, society might benefit from paying closer attention to such technologies. First, while it can feel disempowering to realize how much we rely on technology and on other people, this awareness can help us properly value our interdependence. Second, citizens in a democracy who understand the critical role of often invisible technologies may be more inclined to support government efforts to maintain infrastructure and regulate private enterprise. And finally, simply because it's interesting! It brings richness to the world to understand more clearly how it works.

J. Coupland, *Containing Nature*, https://doi.org/10.1007/978-3-032-11882-0_11

11.2 Reflections for Modern Scientists and Technologists

Whatever the costs to society at large of our unawareness of the technologies that support us, the costs to the people who work in technology are greater. If we believe that inventing new things is what is really important, then the people who maintain and improve existing things—that is to say most technology workers—are not. In this light, the contributions of the many technology workers who spend their careers making incremental advancements are not only meaningful—they are essential.

Technology workers themselves perpetuate this misconception in the ways they talk about their professions. In the first chapter, I called particular attention to the way food scientists, led by Katherine Bitting, have followed this pattern by celebrating Nicolas Appert as the founder and hero of their profession. That isn't to say Nicolas Appert, that old revolutionary, doesn't deserve recognition. He took things that already existed and, over the course of decades, adapted them to reliably preserve food. But, if we look at his invention as the only important thing, we are missing the real ways that technology develops and why technology work is so essential. The real reason Appert is so important is because he was copied.

Once the possibilities of Appert's invention were recognized, people adapted it to meet their local needs. The cycles of copying, adapting, and selecting technologies gave rise to a variety of different ways food could be preserved. The more successful adaptations spread across the industry and became standards, while other adaptations found use in niche applications or simply died out as they were replaced by alternatives. If evolution rather than revolution is the better metaphor for technological change, then, rather than looking to the inventor as the hero for the profession, modern technology workers should look to the people who made or selected these adaptations. What are the qualities or virtues evident in their work which might inspire us to do better today?

Bryan Donkin and Peter de Girard, the engineers mainly responsible for adapting Appert's bottling process to metal cans, and Charles Ams and his team, the inventors of the sanitary can, all made major improvement to the technical processes of canning. They brought their own skills to an existing process and made changes that were eventually widely adopted. This required both imagination, i.e., to conceive of a new way of doing things, and courage, i.e., to implement it. However, imagination and courage can quickly lead to disaster without the judgment to decide on the best course of action to pursue.

The Hume brothers, who traveled west with Andrew Hapgood to start a salmon cannery in California; Edwin Norton, who built the first integrated can-making factory in Chicago and later the Continental Can Company; and Jack Dorrance, who led the Campbell's Soup Company, all ran businesses and used their best judgement to try to organize technology and labor to meet their local conditions. When they made wise decisions, their businesses prospered. Sound judgment is strengthened by ongoing engagement with the practical realities of the business—like Appert's meticulous notes on how to properly drive a cork into a bottle or Campbell's executives participating in daily soup tastings. It also depends on a wide professional network for exchanging ideas, such as those fostered through the meetings and journals of regional canners' associations. Unfortunately, communities can also reinforce harmful beliefs, for example, the widespread racist assumptions about Chinese cannery workers and the near-universal enthusiasm among business leaders for Bedaux's questionable management methods.

There are some additional virtues more relevant to scientists interested in working on technology problems. First and foremost is their responsibility to do good science. Both James Esty, the microbiologist who painstakingly measured the thermal destruction of hundreds of strains of *Clostridium botulinum*, and Frederick vom Saal, the Missouri biochemist who pioneered the meticulous studies needed to show BPA could cause harm across generations of animals, were respected laboratory scientists. They were not "pure scientists" working on some abstract problem with no obvious connection to the real world. They knew their results were important to achieving a goal they cared about, but they recognized that, as scientists, the most important thing they could do was get their facts right. They paid careful attention to their methods and generated the best data they could.

However, even with the best data, science doesn't offer a timeless truth, but rather a process for trying to reach a better understanding of the natural world. Everything we learn is conditional and subject to revision in the future. Robert Boyle and Joseph Louis Gay-Lussac were doing good science when they clearly set out a theory for food spoilage—even if that theory was later proved wrong. In contrast, Justus von Liebig and Friedrich Wöhler were failing as scientists when they mocked rather than engaged with the growing evidence of the importance of microbial fermentations. We are all attached to the belief we are in the right, but science is one of the few reliable ways we have of proving ourselves wrong. Scientists should celebrate corrections and revisions and perhaps be less confident that what we know now is the whole truth.

The communities and networks within science are important. By organizing the disagreement between Louis Pasteur and Félix Pouchet as a formal contest, the Academy of Sciences in Paris was able to settle the question of spontaneous generation. Forming a diverse Botulinum Commission allowed the NCA to find consensus on the regulation of canning, while the Wingspread group helped turn vom Saal into an effective campaigner. Figures such as Charles Bigelow and Theo Coburn, people who can mobilize scientists toward an end, play a distinct yet essential role, and their skills should be celebrated. However, we should also note that organizing science toward a specific goal can easily come into conflict with the other scientific virtues of valuing good data and being willing to change your position in response. This is a special concern for scientists engaged in political pursuits.

Lastly, being effective in applied science also requires getting outside these scientific communities. Harry Russell and Sam Prescott left their universities and went out into the canneries to learn what was being done there. They conducted scientific experiments using industrial equipment. They valued what the canners knew and used it to design experiments and solve problems. This is not applied science as a "one way" transmission from the university to the factory, but a respectful, two-way sharing of scientific and practical knowledge.

Rather than looking for a single "hero of the profession," it is more helpful for technology and science workers to recognize and cultivate "virtues of the profession" such as these. Inventing something as novel and important as canning is beyond the hopes of most of us, but the work we do is important, and good examples can help us do it better.

Finally, there really is no "finally." As I complete this work in the summer of 2025, there are new papers in the latest issues of food science journals describing the importance of calcium on the texture of beans canned in flexible pouches and the development of a bioreactor for utilizing the wastewater from vegetable canning. Another new paper describes the toxic effects of BPA on freshwater shrimp, while Sherwin-Williams is promoting a can liner made without it. JBT, a canning machinery manufacturer,[1] has designed a novel way to circulate fluids inside a retort, so cans at the center of a stack heat at the same rate as those on the outside. Campbell's Soup is investing in their North Carolina plant where they sterilize their soup before adding it to the containers to preserve a better flavor and more nutrients; the Dorrance family is still the largest shareholder in the company. Trident Seafoods, a major Alaskan salmon canning company, is partnering with a college in Washington to develop a 12-week intensive course to train workers to fix equipment in the remote canneries where parts can be hard to source. The Consumer Brands Association[2] is campaigning against tariffs on the tin-plated steel used to make cans and offering courses on labeling regulations for food manufacturers. The invention of canned food continues.

Not all of these innovations will prove to be good ideas. Indeed, many "solutions" which are adopted lead to unforeseen problems: the shift from glass bottles to soldered cans improved the process but exposed consumers to lead; flexible pouches are lighter and safer for military rations, but they are less recyclable than metal cans; the development of epoxy resins gave better protection to the metal but released BPA into the food. Nevertheless, a characteristic virtue shared by the technology workers in this book was a discontentment with how things were and a belief that improvement was possible—if only they could build a better "active human interface with the material world."[1]

[1] JBT has roots in can making that reach back to the 1920s and incorporates the FMC canning business where Charles Stumbo worked in the 1940s.

[2] A trade organization that developed from the National Canners Association which, under the leadership of Willard Bigelow, had directed the efforts against botulism in the 1920s.